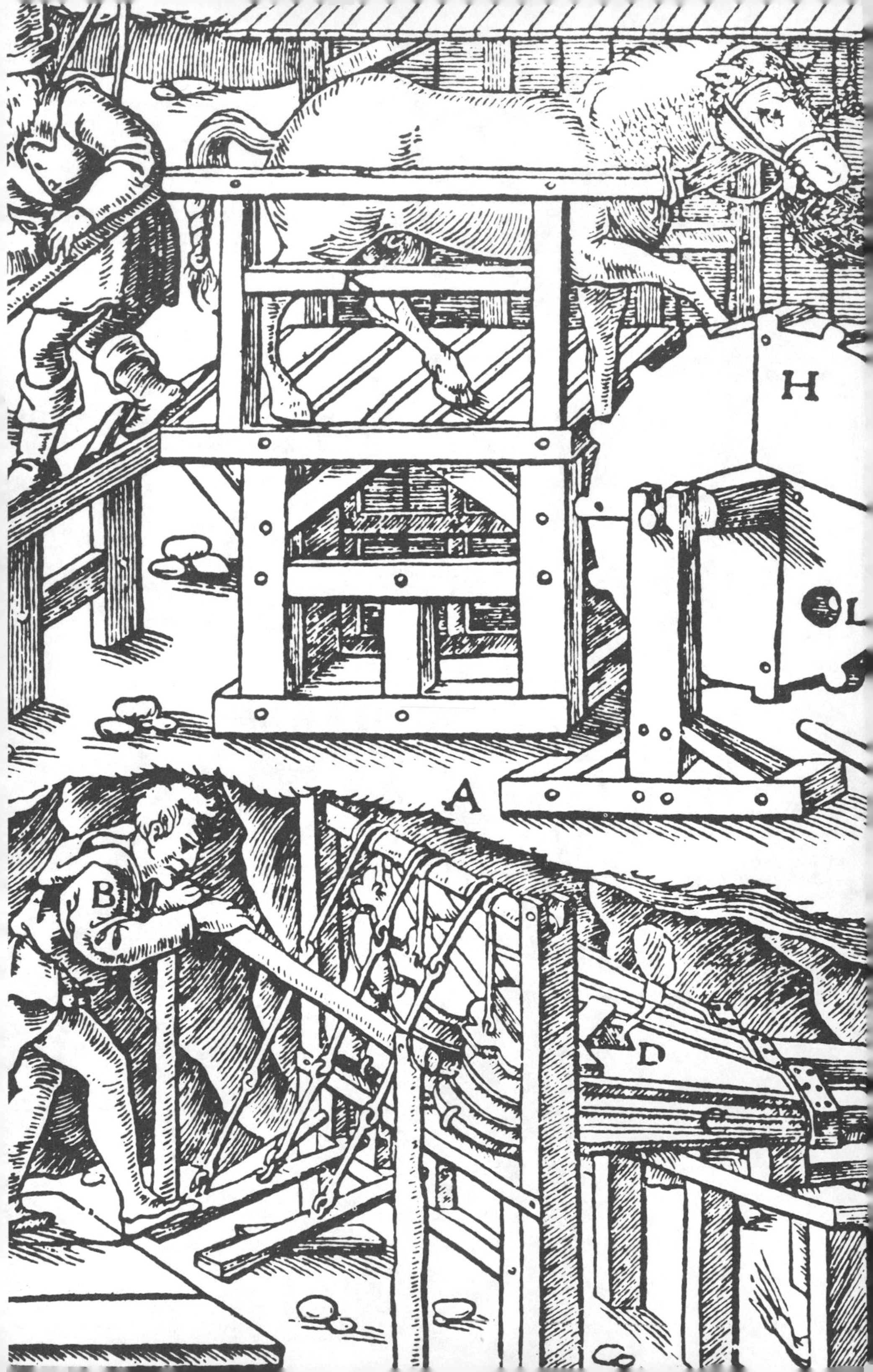
H
L
A
B
D

DRY WASHING FOR GOLD

For Jim
From Dad
6-20-96

JAMES KLEIN

GEM GUIDES BOOK CO.

For Jane,
someone who cares,
with love.

Published by:
Gem Guides Book company
315 Cloverleaf Drive, Suite F
Baldwin Park, CA 91706

Cover by Lynn Van Dam.

Library of Congress Card Number 94-77565

ISBN 0-935182-76-4

Printed in the United States of America.

Contents

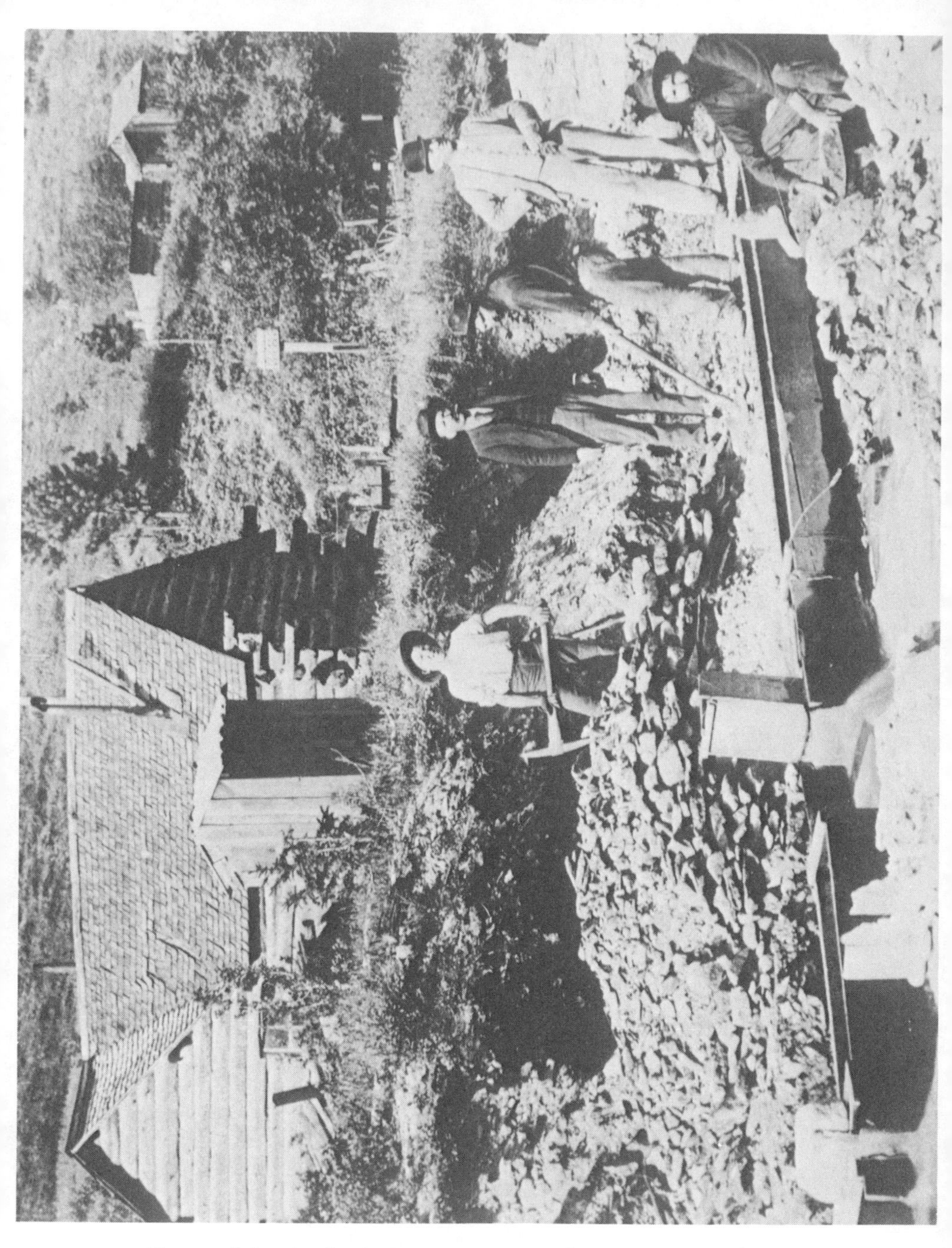

Placer mining in the 1800s. Courtesy Utah State Historical Society

Introduction

Anyone who has spent any time prospecting for gold knows the value of a good dry washer. In my mind there is no more important piece of equipment for the small scale gold miner. The reason for this is the fact that the greatest untapped source of placer gold lies not in the present stream beds but in the more inaccessible areas. I am referring to the bench deposits and terrace gravels in the canyons cut by the present streams, the dry stream beds in the deserts and arid regions of the world, and the ancient stream beds left on the tops of mountains or buried by geological upheavals in the past. I hope this book will help you in your search. In it you will learn where to dry wash and the best methods available for the recovery of the gold.

Since I wrote the first edition of *Dry Washing For Gold*, there have been great advances in metal detector technology, so I am including a chapter on prospecting with a metal detector in this new edition.

There have been several outstanding finds made with the metal detectors recently, such as the *Hand of Faith* nugget worth over $1,100,000, a large heart-shaped nugget found in the Mojave Desert, and thousands of smaller nuggets as well.

Dry washers are still basically the same construction and the gold-bearing areas haven't changed, so I have left things pretty much the same.

There's still a lot of gold out there and it's just waiting to be found. I hope everyone of you finds El Dorado. I know you're going to enjoy the hunt . . . .

Jim Klein

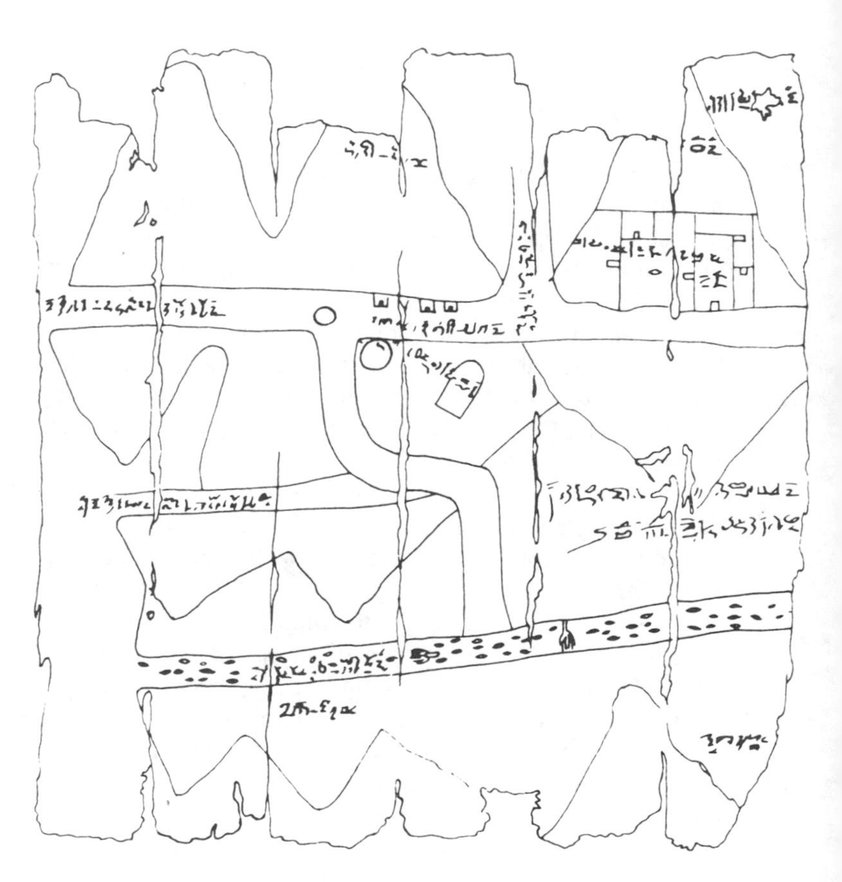

Ancient Egyptian map, showing location of a gold mine.

Chapter 1
The History of Gold

Man's fascination with gold goes back to the days when we still lived in caves. Scientists have discovered the first man-made objects of gold in caves that date back to Cro-Magnon man in Europe. Primitive man was the first to associate its rich warm yellow color with the sun. The first objects made to represent the sun were hammered out of nuggets found in the stream beds. References to gold appear in the earliest writings of the Bible and in ancient Greek legends.

The first recorded mining is known to have taken place in Egypt where the placer deposits of the Nile River are thought to have yielded over a million pounds of gold. The Egyptians were also the first to use beaten gold in their works of art. Another first by the Egyptians was the smelting of gold. The art of these early goldsmiths was incredible, as we can see in the artifacts recovered from the tombs by archaeologists.

The Egyptians worshipped Ra, the Sun God, whose hawk-like image is found in many artifacts. They felt that to possess one of these sacred golden images gave powers not held by mortal men. They felt this so strongly that the kings and rulers of these times had their tombs bedecked with golden objects so that they could take this power and influence with them into the land of the dead. We can only guess at the amount of gold that was buried with the kings, since the tombs were victims of tomb robbers almost as soon as they were sealed. One look at the treasures unearthed from the tomb of so young a king as Tutankhamen boggles the mind, and the treasures that were lost must be beyond compare.

The Egyptians also mined the lode deposits in Nubia. The kings would send the condemned criminals as well as war captives to work the mines. Sex or age made no difference to them. Young boys were used to pick up the pieces of ore broken off by the miners and take them to the place where the women and older men would crush it.

The Bible tells us on the second page of Genesis that there was gold in the land of Havilah. We have all heard of King Solomon's fabulous mines. So much gold was mined by Solomon that his temple and its furnishings were covered with it. His mines were thought to be located on Mt. Sinai, recently thought to have been rediscovered. We also read of King Croesus of Lydia, who gained favor with the Greeks by sending 7,500 pounds of gold to the shrine of Apollo at Delphi. In early Greek literature we hear of the golden calf, King Midas's golden touch and of the Golden Fleece. The golden fleece is thought to be a sheepskin used in a sluice box that became so full of gold that it seemed to be pure gold.

The Romans found new ways to increase production from the mines. While mining the lode deposits in what was then Gaul, they developed a method of heating the rock, then throwing cold water on it, causing it to crack and shatter. They were also the first to use hydraulics to mine gold.

During the Middle Ages, alchemy became the rage, as men sought the so-called "philosopher's stone" in order to turn base metals into gold. Alchemy wasn't new; its history can be traced back to ancient China. The Chinese believed that gold possessed special powers and could even prolong life.

A scientist who lived around 750 A. D. named Gerber searched all his life for the stone, and to him goes the credit for the discovery of nitric acid, which we use to clean gold, and also for the discovery of nitrate of silver.

When workers were tearing down a building in the town of Kufa, Iraq, around the year 1000, they discovered Gerber's Laboratory. According to reports it was just as it was when Gerber died. It had more furnaces than a modern laboratory. The furnaces were all shapes and sizes, and were used for the smelting of ores in his search for the philosopher's stone. Everything he used was in perfect condition. There were flasks, vials, ladles, beakers, tongs, molds, crucibles, rock crushing equipment, and metal rollers. Even his notebooks were there. Gerber lived into his 90s and was one of the most famous scientists of the Middle Ages.

He never discovered the formula for the philosopher's stone. It was not to be discovered until much later, 1936 in fact. It was in a cyclotron at the University of California that gold was transmuted from another metal by nuclear reaction. Unfortunately, the metal used is even more rare and costly than gold; it was platinum.

Alchemy is an Arabic word. To the Chinese it was *Kem-mai* which means, "gone astray in the search for gold". To the Greeks it was *chemeia,* which means "black," as they felt it was one of the black arts. The "AU" symbol used by scientists today originated in the eighteenth century. It is an abbreviation of the Latin word for gold, *aurum*. The word aurum comes from Aurora, the goddess of the dawn.

When Caesar captured Gaul, he had a large network of roads built to transport his booty back to Rome. The gold was collected by tribute collectors and carried in sealed baskets called "fisci," after being stored in a temple named the "fiscus." This is where the word fiscal originated. Roman coins called "aurei" were minted in the temple of the goddess Juno Moneta. It is from her name that we received the word "money."

Other famous alchemists of the Middle Ages were men like Jildaki of Cairo, who proposed what is now known as atomic structure. The Englishman Roger Bacon, a teacher at Oxford, made great advances in the science of optics, and also predicted the airplane and the steamship. The Swiss alchemist, Paracelsus, while working with quicksilver, discovered a remedy for syphilis. Basil Valentine developed crystalline antimony while trying to make gold. Even Isaac Newton dabbled with the transmutation of gold. Johann Bottger, a genius in alchemy, was locked up and guarded all his life by one ruler or another in the hope that he would unlock the door to the secret of the philosopher's stone. While being kept in the Gold House in Dresden, Bottger developed a method of making a porcelain so fine that it became famous as Dresden china.

It was during the 1500s that the art of goldsmithing reached its peak. Men like Benvenuto Cellini, Michelangelo, and Leonardo Da Vinci all began as apprentice goldsmiths. Some of the most beautiful works of art remaining today came from the hands of men like these.

While the Old World was searching for a means of increasing the amount of gold in their treasure houses by any and all means, the Indians of the New World had been mining the great riches there since before the time of Christ.

It was a poor sailor who had a dream that you could reach the rich ports of Cathay in the east by sailing to the west. He would be the one to open the way to the new land. For years he could find no one to finance his plan. He presented his idea to almost all the courts in Europe, but to no avail. At last, one of the poorest of them all, Spain, decided to take a chance on him.

Thus it was that Christopher Columbus, in his desire to open a new route to the golden treasures of the East, should discover one of the greatest storehouses of gold ever known to man. He was followed by Pizarro in Peru, Cortez in Mexico, and Sir Francis Drake along the west coast of America. Spain became the richest nation in Europe when in just 50 years (1550-1600), she took over 2 million pounds of gold from the captive land.

On the island in the Bahamas where Columbus first touched down, he found the natives wearing gold rings in their noses and fishing with gold hooks. When he returned to Spain with the new-found gold, he reported that he had found a land rich in gold mines and that the rivers were filled with gold. He became an instant hero and set off a rush by Spaniards to be the first to harvest the golden crop.

The first Spanish fleet to arrive in the Indies was made up of sailors, rich men, poor men, farmers, priests, and adventurers. They found nothing but hardship, mud huts, and hostile natives. In 1500 Columbus was returned to Spain in chains. The chains were removed when he got back and the charges against him were dropped, but he never forgot what they did to him, and always kept the chains close by to remind him for the rest of his life.

In 1517, Hernandez De Cordoba and his crew sailed south from Cuba and became lost in a storm and found themselves in strange waters near a shoreline unknown to them. They were able to see stone walls and the tops of great pyramids behind them. They thought they were in India and named the city Cairo. They landed and advanced toward the city and were met by native warriors who attacked them with lances and arrows. They frightened them off by firing their muskets and were able to capture one of the closest temples. Inside the pyramid were idols and images made of gold.

They carried off all they could find and set sail down the coast looking for other cities to plunder. They were not as fortunate when they attacked a second city. The Indians were not frightened by the musket fire and managed to kill and capture many of the invaders while wounding the rest. Cordoba died from the wounds he received

Deserted gold mine. Courtesy Western Treasure Magazine.

from this battle, but not before he was able to return to Cuba and spread the word of his discovery. The Mayan Empire had eluded the Spanish for almost a quarter of a century since Columbus had first sailed into these waters. The Mayan civilization was as old as the Egyptian and maybe even older.

Hernando Cortez was in Cuba when Cordoba returned. When he heard of Cordoba's discovery, he slipped out of the harbor with his ship and crew in search of the gold. The rape of the Americas was on. After Cortez captured Tenochtitlan (the site of Mexico City), his men spent three days melting down all the gold objects into bars. It came to almost two thousand pounds. Much of Cortez's loot was lost when he was attacked by the Aztecs and had to flee the city. Many of his men drowned, weighed down by stolen gold in their retreat. Later, when he returned, he could find no trace of the gold left behind.

As rich as Montezuma's treasure was to Cortez, it was not as great as Pizarro's. When Pizarro tricked the Inca ruler, Ataualipa, into being captured by a small force of Spanish (62 horsemen and 106 infantrymen), the Sapa Inca offered to fill a room seventeen feet wide by twenty-two long with gold as ransom. It took nearly two months for the gold to be collected. As a reward for paying his ransom, Ataualipa was strangled. The gold in the room weighed thirteen thousand pounds.

In 1577, Sir Francis Drake set out on a trip around the world. Drake was a navy officer to the English, but to the Spanish he was a pirate. During his trip he looted several Spanish ships. When he reached the coast of California, his ship, The *Golden Hind*, was laden with 20 or 30 tons of gold. He became becalmed near what is now San Francisco Bay. While there he claimed California for England and left a plaque commemorating the event nailed to a tree. It is believed this plaque has just recently been found. Drake, in his search for gold, never knew he was only a few miles from the Mother Lode.

The first miners in America came from Europe. As early as 1816, a book on mining published in England told of great amounts of gold waiting to be mined in the mountains of California. The Padres from Spain were aware that there was gold to be had there early in the days of the missions in the 1700s. The Indians would bring gold to the mission in exchange for items they desired. It is said that several of the missions had their own gold mines worked with Indian labor.

There are several lost treasure stories that have been handed down to modern times connected to these mines. I've even spent some time of my own searching for some, such as the Lost Padre, said to have been located in the mountains north of Los Angeles.

The big story of gold in America is the gold rush of 1849 in California. It was James Marshall building a saw mill for John A. Sutter on the American River, at a place known as Coloma, who found the nugget that started it all. The California gold rush was unique, as it was the first time in history that the gold mined went to the miner, not to a ruler, nor was there any tribute to be paid.

Actually, at the time of Marshall's discovery, California still was a part of Mexico. He found the gold on January 24, 1848. The peace treaty of Guadalupe Hidalgo, which gave California to the United States, was signed on February 2, 1848. So for the first year, the miners made their own laws, held their own courts, and handed out their own brand of punishment to anyone who broke their laws. Since there was so much gold, there were not too many problems at

James Marshall

first. It was only later, when those who wanted to live off the labors of others arrived, that there was any real crime. The law in the mining camps varied. When a strike was made, the miners would

Old building in gold camp, now deserted. Photo by Larry Winkelman

come together and establish the rules and regulations of that camp. The laws dealt with the size of a claim a man could own and the number of claims he was allowed. They felt that a man should only have one claim and were against a man having slaves or any other form of labor working a claim. It was really a true democracy, with every man for himself and equal rights to all. The worst crime was theft. Anyone found guilty of stealing any tools or supplies from another man's claim was often hung on the spot. Other crimes like trespassing meant banishment and branding.

When Marshall made his discovery, there were only about 14,000 people in all of California. By summer, most of them were in the mountains. The S.S. *California* arrived in San Francisco on February 28, 1849, carrying the first argonauts from the east, and the rush was on. Within a few months there were 100,000 men in the mountains. The California gold rush is said to have lasted until 1855. It was the richest gold rush of all time for the individual miner. The average miner was able to pan at least an ounce of gold per day. Gold was found from Mariposa in the south to Downieville in the north, an area over two hundred miles long in the foothills of the Sierra Nevada.

Washing gold from the blankets. Courtesy California Division of Mines

The Mexican Arrastra. Courtesy California Division of Mines

Neither John Sutter or James Marshall profited greatly from the gold. Sutter suffered the loss of the large holdings he had acquired, as well as his dream of being ruler of a vast domain. He sued the United States government for the losses he suffered at the hands of the 49ers who had invaded his land, and died in 1880 in Washington D. C., still trying to get justice done.

Marshall wandered the foothills for years in search of a rich strike, but it always eluded him. He did seem to have an ability to locate rich deposits, but not for himself. Miners would follow him

Crushing the ore. Courtesy California Division of Mines

and work an area after he would leave it; several rich finds were made this way. For instance, shortly after his discovery at Sutter's Mill he tried his luck in the area of Deer Creek near Nevada City without success. A year later, 5,000 miners were at work in the area, which produced $378 million over the next hundred years. Marshall died broke, a bitter man, in a small cabin in Kelsey, near Coloma.

By now, the whole world was infected with gold fever and every report of gold being discovered set off a rush to the area. Thirty thousand miners left California in 1858 and rushed to the Fraser River in British Columbia when word reached them that gold had been found there.

Next it was to Nevada, and the fabulous Comstock Lode. Many Forty-Niners had passed over this area in their rush to the Mother Lode country. Ten years later, they were back digging the rich gold and silver ore. The Comstock became famous as a silver-producing area, but the ore contained 43 percent gold, and it was the gold that actually produced the greater profits during its boom.

Also in 1859 there was a rush to Colorado. "Pike's Peak or Bust" was its motto—and it was mostly bust. The miners did better a few months later at Gregory Gulch, also in Colorado. Here they mined the so-called "Blossom Rock," quartz outcroppings that could be easily crushed and then panned or sluiced. The summer of 1859 found 10,000 men working in the four-square-mile area.

America was not the only place that was a Mecca to gold seekers in the 1850s. Large strikes had been made in western and eastern Australia at the same time; men were flocking to those gold fields. One of the unique things about gold in Australia is the number of large nuggets found there. In 1851, a man named Holtermann found a mass of gold and quartz that was four feet nine inches high and two feet two inches wide containing 3,000 ounces of gold. They are still finding large nuggets there today.

Edward Hargraves was the James Marshall of Australia, having found the first gold on the Macquarie River in 1851. He left Australia in 1849 to join the rush to California. He didn't strike it rich in the Mother Lode, but he did learn about mining and what gold country looked like; it looked like home. He returned to Australia, made his discovery and was rewarded by being made Commissioner of Crown Lands—a better fate than Marshall's.

In the 1860s and '70s, gold was found many places in western America. Strikes were made in such places as the Black Hills in South Dakota in 1865, in New Mexico a little later, and then came the rich discoveries at Cripple Creek in Colorado.

The last great "free man's" gold rush was to the Klondike in 1896. The Klondike gold fields were found by an old-time prospector named George Washington Carmack. He was led to the placer deposits on a river by some Indian friends. The first year, the average *day's* wages by the miners was $866. Since it was such a remote area, news of the new strike was slow in spreading It was not until July 14, 1897, when the steamship *Excelsior* docked in San Francisco bringing 40 prospectors and a half-a-million dollars in gold ashore, that the world first heard the news.

There were two ways to get to the gold fields. If you had the money, you could take a ship to St. Michael and from there another ship would take you up the Yukon River all the way to the gold fields. Most men didn't have the exorbitant fares and were forced to take a more difficult route. They would book passage to Juneau or Skagway and then backpack 600 miles over Chilkoot or White Pass. It is said that during the winter of 1897 thirty thousand cheechakos, or greenhorns, made the journey over the passes. It was so difficult in places that travelers could only go single file. At times they had to crawl on their hands and knees since the trail was so steep. Sometimes men waited for days for an opening in the line of gold seekers filing over the pass.

Landslides claimed many of the miners. A landslide on White Pass killed 63 men in April of 1898. The snow was so heavy that only seven bodies were discovered. Even with all the hardships 5,000 pounds of gold was mined in 1899, and 45,000 pounds of gold was taken out of the tundra in 1900.

For centuries, men had traded along the East African coast for gold, but nothing was known of the mysterious jungle interior. Then, in 1868, explorers in Southern Rhodesia discovered the ancient site of Zimbabwe, which was made up of hundreds of stone fortresses, each guarding an ancient gold mine. There was a palace and a temple enclosed by walls 12 feet thick. Gold artifacts were found in the ruins of the temple and palace as well as in the thatched huts surrounding the ruins.

In 1889, Cecil John Rhodes, an Englishman, led a large party of prospectors and guards to the area and claimed it for Great Britain. This was the beginning of the huge mining operations of South Africa.

As many as 400,000 black workers worked the mines, usually managed by whites. Seventy-five million metric tons of ore are mined every year. Each 100,000 tons of ore produces about a ton of gold. The shafts are as deep as two miles into the ground. Since 1890, over 800 million ounces of gold has come out of these mines. South Africa is the largest producer of gold in the world today. The Soviet Union is the second largest producer, followed by Canada, then the United States, Central and South America, Ghana, New Guinea and Australia.

For years there had been only one price for gold, which was set by world agreement. Then in 1968, the major nations established a dual system. There was now an official price for gold trading among nations and a free market price. The free market price is based on supply and demand. The official price remained at the $35-per-ounce figure established in 1933. The open market jumped immediately to $40. In 1969 the market price dropped back to $35, and in 1970 it fell below that to $34.85. By 1971 it had risen back up to $42. Then, in May of 1972, it hit $58. A year later it topped the $100 mark. In November of 1973, the dual price system was dropped and governments could sell their gold on the open market.

In 1975 the United States lifted its ban on private ownership of gold and by early 1979 the price of gold had jumped to around $300. At one point in 1980, the price reached $800 per ounce, but then fell back as profit-taking hit the market, but only to around the $500 mark. This high price and the knowledge that it will rise again, (I would not be surprised to see it hit $1,000) has prompted renewed gold prospecting. Now almost anyone with the proper equipment and a little knowledge can make money looking for gold.

One thing to keep in mind about gold is that it always keeps its value. A $20 gold piece a hundred years ago would feed a family for a long time and a $20 gold piece today will still feed a family for a long time. Now, what about that $20 dollar bill?

139. Mining Methods of Early Days

Early mining methods included a tunnel, rocker, pan, sluice, shaft, long tom and coyote hole. (Photo courtesy of Security Pacific National Bank)

Chapter 2
Geology of Placer Deposits

In order to have the best chance at finding that golden dream you are seeking, you need to have some knowledge of placer deposits. A lot of our information comes from the early miners and prospectors who climbed, dug into and checked every mountain, canyon, stream, river and creek looking for gold. This is still the best method, as geologists admit that even today they don't know everything there is to know about the remaining rich gravel deposits.

In a *Service Bulletin* put out by the California Division of Mines and Geology, it is stated that "The geologic history and structure of the buried channels are so complex that the best of engineers have been baffled by them. Fragmentary benches and segments of rich gravel deposits which still rest in positions completely hidden from the surface, or even from the underground passages which enter into the lower main channels afford alluring possibilities to the geologist as well as the prospector." They are telling us that there is still a lot of gold out there; you have as good a chance of finding it as any geologist.

Keep in mind that most areas have been prospected at one time or another. Don't waste a lot of time in areas that have not proven to be productive in the past. Search the areas that are known to be gold-bearing and take advantage of the knowledge gained by those who went before you.

There are several types of placer deposits which are classified here to indicate how they were first formed. The basic placers are:

(1) Residual placers or "Seam Diggings."
(2) Eluvial or hillside placers, representing transitional creep from residual deposits to stream gravels.
(3) Bajada placers, a name given to a peculiar type of desert or dry placers.

(4) Stream placers, which have been sorted and resorted, and are simple and well-merged.
(5) Glacial-stream placers, which are for the most part profitless.
(6) Eolian placers, or local concentrations caused by the removal of lighter materials by the wind.
(7) Marine or beach placers.

Of the seven types, the stream placers are the most important; they have been the source of most of the placer gold mined in California. Stream placers consist of sands and gravels sorted by the action of running water. If they have undergone several periods of erosion, and have been resorted, the concentration of heavier minerals is greater.

Deposits by streams include those of both present and ancient times, whether they form well-defined channels or are left merely as benches. All bench placers, when first laid down, were stream placers similar to those of present stream deposits. If not reworked by later erosion, they are left as terraces or benches on the sides of the valley cut by the present stream. These deposits are called bench

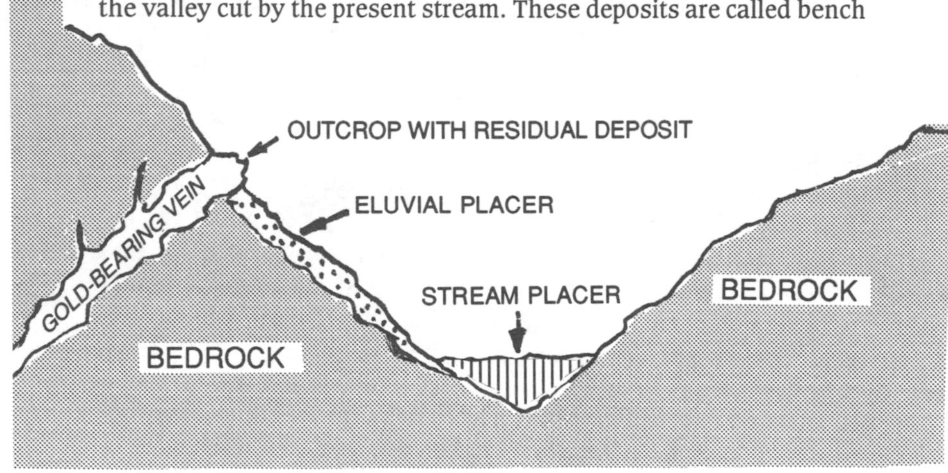

Most gold comes from intrusive veins, which break up to form eluvial deposits. These deposits move downhill until captured by moving water, which continues the separation of gold from its original rock matrix.

gravels. In order to understand stream placers, streams themselves must be studied in regard to their habit, history and character.

Residual placers are formed when gold is released from its source and the encasing material broken down by long, continued surface weathering. Disintegration is accomplished by persistent and powerful geologic conditions which affect the mechanical breakdown of the rock and chemical decay of the minerals. The surface of a gold-bearing ore body is enriched during this process of rock disintegration, because some of the softer and more soluble parts of the rock are carried away by erosion.

After gold is released from its bedrock encasement by rock decay and weathering, it begins to creep down hillsides and may be washed down rivulets and gullies into stream beds. On its way down the hillside, gold is sometimes concentrated in sufficient value to warrant mining. These concentrations are classified as "eluvial" deposits.

It is a common fallacy of some prospectors to attribute the forming of some placer deposits to the action of glaciers. Since it is the habit of glaciers to scrape off loose soil and debris but not to sort it, ice is ineffective in the concentration of metals. The streams issuing from the melting ice may sometimes be effective enough in sorting to create a deposit.

Bajada is a Spanish word for slope, used to identify a confluent alluvial fan along the base of a mountain range. The total production of gold from bajada placers is small compared to other placer workings, due to the less-efficient dry washing methods used in the past. The forming of a bajada placer is basically similar to a stream placer, except as that it is conditioned by the climate and topography of the arid region in which it occurs. The bulk of the gold that has been released from its matrix as it travels from the lode outcrop to the bajada slope is deposited on the slope close to the mountain range. The gold is dropped along the lag line, which is the contact of the basin fill with the bedrock. Most eolian placers of the desert are as a result of the bajada being enriched on the surface by wind action on the lighter materials.

Although the heaviest concentration of gold will be on bedrock, bulk concentration does not occur as in a stream deposit. Since a certain percentage of gold is still locked in its matrix, there is a strong tendency for less gold to reach bedrock and for more to remain distributed throughout the deposit than in the case of stream gravels.

There have been several beach placers found and worked along the Pacific Coast. Beach placers result from shore currents and wave action on the materials broken down from the sea cliffs or washed into the sea by streams. There are two types of beach placers: present beaches and ancient beaches. Most gold-bearing beaches are found in northern California. Not a great deal of production has come from these deposits. Most gold will come from rocks that are being eroded by the waves.

There are several things that occur to preserve a placer deposit. Since streams are constantly changing their position, fragments of their deposits are left isolated. For example, the benches and terraces that are left at different intervals when a stream is cutting a deeper channel will eventually be eroded away, unless something protects them.

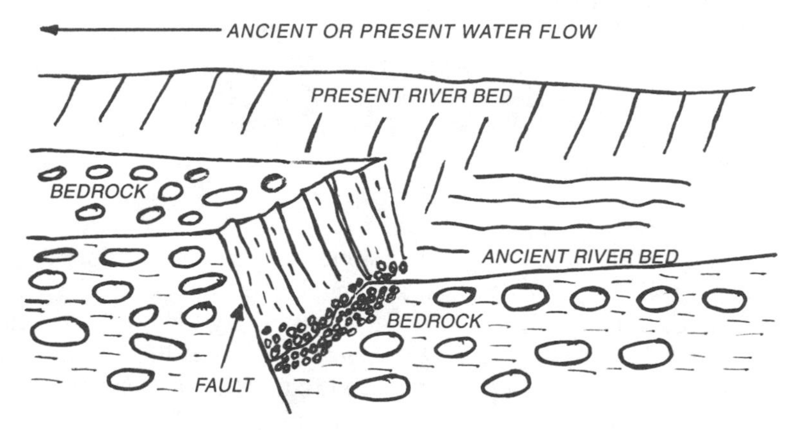

Cross-section of a river that has suffered down-faulting on the upstream side, creating a pocket to trap the gold, gravel, sand and silt.

Burial is the most effective way a placer may be preserved. When the term "buried channel" is given to a placer, it is one where a stream has been covered by lavas, mud flows or ash falls. There are other ways by which placers may be buried, such as:

(1) By covering with landslide material.
(2) By covering with gravel.
(3) By covering with lake deposits.

(4) By covering with gravel when the stream is choked.
(5) By covering with gravel when the stream course is lowered below the general base-level of erosion.
(6) By the covering of older stream courses with alluvial fan material, as conditions favorable to stream existence fail.
(7) By covering with glacial till.
(8) By covering of beach placers with marine sediments as the coast is submerged and elevated.

The gravel content of a placer may become firmly cemented when it is infiltrated by mineral matter, such as lime and iron carbonate or silica. The older the placer, the more likely this is to occur. These cemented gravels are sometimes very hard to break down, which is why some old mine tailings are profitable to work. The cemented gravels sometimes were never completely broken down as they traveled through the sluice boxes, and the gold was redeposited in the present stream bed.

The gold found in placers originally came from veins and other mineralized zones in bedrock when the gold was released from its rock matrix by weathering and disintegration. Many times, the source of the gold in a placer would not be a deposit that could be mined at a profit, but the richer deposits usually indicate a comparatively rich source. Sometimes a rich placer will be developed when several low-grade veins feed it over a very long time. The richest placers are created when there is reconcentration from older gold-bearing gravels. For the most part, the original source of the gold is not far from the place where it was first deposited after being carried by running water.

Mineral grains that are very heavy and resistant to mechanical and chemical destruction will be found with the gold in placer deposits. These are what prospectors and miners call black sand. Black sand is principally magnetite, but some other minerals you will find in your sluice box are garnet, zircon, hematite, pyrite (fool's gold), chromite, platinum, cinnabar, tungsten minerals, titanium minerals and possibly even diamonds. You'll find a lot of other things, such as quicksilver, metallic copper, amalgam, nails, buckshot, BB's, bullets, and what-have-you.

The very high specific gravity of gold, six or seven times that of quartz (increasing to nine times under water), is what causes the gold to work its way down to bedrock or false bedrock, or any point where it can go no farther. Once gold is trapped on bedrock, the stream has great difficulty picking it up again. The specific gravity of gold is 19; that is, it weighs 19 times as much as an equal amount of water to its mass.

AVERAGE SPECIFIC GRAVITY OF SOME MINERALS

MICA	2.3
FELDSPAR	2.5
QUARTZ	2.7
HORNBLENDE	3.2
GARNET	3.5
CORUNDUM	4.0
MAGNETITE	5.2
SILVER	7.5
GOLD	19.2

Due to their insolubility, the finest particles of gold are preserved. A piece of gold worth less than a dollar can easily be recognized in a pan. Since gold is so malleable, it will be hammered into different shapes by stones as they tumble along in the stream. It will not be welded together to form larger nuggets, as some people believe. Particles of gold may be broken off, however, from another piece.

Geologists have shown that the largest masses of gold come from lodes and not placers. The more rounded and flattened nuggets have probably been in the stream for a longer time and have taken more knocking-around than the ones that show their original crystalline form. Crystalline nuggets are known as coarse gold and probably have not traveled far from their source.

Gold found in the more ancient placers has a higher degree of fineness than that whose source is nearby. This may be due to the removal of alloyed silver by the dissolving action of the water.

The accumulation of gold in an important placer deposit is not just pure coincidence, but is the result of some fortunate circumstances. In areas where nature has provided extensive mineralization, rapid rock decay and well-developed stream patterns, there is the opportunity for numerous gold placers to be formed.

Basically, what happens is fairly simple: In areas where the gold has been deposited, the power of the stream has become insufficient to carry off the particles of gold that have settled. The richness of the deposit depends upon how quickly the stream loses velocity, as well as the ability of the bedrock to hold the deposited gold, plus the relationship of gold sources to the stream. When a stream is eroding, materials are constantly moved downstream.

During this movement, a constant sorting is taking place, causing a concentration of the heavier particles. Deposition then takes place in the stream when the velocity is decreased, either by changes in volume or grade. When this happens, the gold is laid down with the other sediments. Sometimes the placer gold may be trapped in irregularities in the bedrock, without considerable detrital material being trapped with it, but extensive placers, as a rule, are not formed by irregularities in the bedrock alone.

When the bed of the stream is the actual floor of the valley, this is true bedrock. However, when the gravels become covered with volcanic or other materials, the stream will flow over this new floor, making deposits on what is known as false bedrock. An area may, therefore, contain two or more layers of gold-bearing gravels. An easy way to see how a stream lays down these various layers is to study areas where road cuts have exposed ancient stream deposits, and also in canyons where benches can be seen.

A smooth, hard bedrock is a very poor location to develop a good placer deposit. Bedrock formations that are highly decomposed and possess cracks and crevices are good; those of a clayey or schistose nature are rated excellent in their ability to trap particles of gold.

Here are some figures from a report by the California Division of Mines and Geology on the size of material carried by a stream flowing at different velocities:

3 in. per sec.	0.17 m.p.h.	will just begin to work on fine clay.
6 in. per sec.	0.34 m.p.h.	will lift fine sand.
8 in. per sec.	0.455 m.p.h.	will lift sand as coarse as linseed.
10 in. per sec.	0.5 m.p.h.	will lift gravel the size of peas.
12 in. per sec.	0.682 m.p.h.	will sweep along gravel the size of beans.
24 in. per sec.	1.364 m.p.h.	will roll along rounded pebbles 1 inch in diameter.
3 ft. per sec.	2.045 m.p.h.	will sweep along slippery angular stones the size of hens' eggs.

As far as grade is concerned, a grade ranging from 30 to approximately 100 feet per mile will favor the deposition of gold. Anywhere that the grade is greater than that, such as in mountain streams or in narrow canyons, will not be a good source of placer deposits.

When a stream leaves its mountain canyons and enters a more level country or a still body of water, the material it carries is deposited in the form of a fan or a delta. At the apex of this fan or delta the fine gold will be deposited, and may never reach bedrock. The steering action that occurs in the rugged mountains during times of floods, which permits gold to reach bedrock, does not take place in the delta.

To sum up, remember that gold is heavier than most material in the stream bed, and it will drop anywhere the flow or grade changes, causing the stream to slow down and lose its carrying power. These are the places you want to search. Each rainy season will bring new gold down from the hillsides into the stream bed.

Keep in mind that the early miners were working deposits that had thousands of years to develop. Try to find material that has not been worked before or try to reach the hard-to-get-to areas where the chances are that the gravels have not been worked as much. The most important thing to remember is that "GOLD IS WHERE YOU FIND IT."

Chapter 3
Where To Dry wash

Now that you have some knowledge of placer deposits, what about the best places to dry wash? I can't count the number of times people have asked me the question: "Where is the best place to dry wash for gold?" The answer is any place that there is placer gold. Most people think that the only area that is good for dry washing is the desert, which is not true. Many good placering regions do not have an abundance of water the year round, but this doesn't mean that you have to limit yourself to working the area only when water is available.

Almost all good prospectors have a good dry washer along with their dredges and sluice boxes. I remember the first time I took my dry washer into the Mother Lode country. A lot of the old boys didn't know what it was until I told them. These men who all their lives had associated gold mining with Long Toms, dredges and sluice boxes would just walk around the dry washer and shake their heads. After they had seen the gold it could produce they all wanted to know where they could get one.

We will discuss the various types of dry washers in another chapter. In this chapter we want to talk about where you want to dry wash.

First, let's list the various locations where gold has been recovered in the United States.

ALABAMA: Gold was found here in the 1830s. During the early days quite a bit of work was done here. Total production is around 50,000 ounces. The gold is found in the counties of Talladega, Cleburne, Randolph, Clay, Coosa, and Chilton.

ALASKA: The last frontier, this area holds one of the greatest attractions for the modern day prospector, the prospect of areas still untouched. The remoteness and shortness of the time available to work the deposits still offer a great opportunity to make good finds. There are over sixty different gold districts in Alaska. Contact the Alaska Division of Mines and Minerals, State Capital Building, Juneau, Alaska 99801, for information and directions to the various locations.

ARIZONA: The earliest known placer mining in Arizona took place about seventy miles west of Tucson in the Quijotoa district around 1774. The most productive areas have been the La Paz placers near the Colorado River 65 miles north of Yuma and the Weaver-Rich Hill placers in southern Yavapai County. Good gold placers are found in all the counties in Arizona except Apache, Coconino, and Navajo.

CALIFORNIA: More gold has come out of California than any other state. It is still producing gold for both the week-end prospector and the professional miner. The gold is found in the mountains and the desert. It can be found in streams and dry washes, on canyon walls and mountain tops.

The Mother Lode region runs for over 200 miles through the foothills of the Sierra Nevada mountains. The southern end is in Madera County and the gold belt continues up to Butte and Plumas Counties. Most of the mountain ranges in Southern California contain varying amounts of gold. Both the High and Low Deserts have gold districts. The mountains of upper Northern California have been very productive, especially the area surrounding the Trinity River. The Coast ranges have also had some production, but it is moderate compared to other areas in the state. The East side of the Sierra Nevada also contains some gold. Even the beaches in some areas have contained gold in workable amounts. There are several good books pinpointing the areas; California's Division of Mines and Geology puts out some good pamphlets.

COLORADO: The placer deposits here have been very productive. The area around Denver and to the west and southwest is where the greatest concentration takes place. There are placer deposits in Jefferson, Gilpin, Park, Boulder, Summit, Lake, Eagle, Routt, Moffat, Clear Creek, San Miguel, Dolores, San Juan and

Costilla Counties. Also, some placers have been found in Hinsdale, Chaffee, Mineral and Montezuma Counties. For more details on the various locations, write to Colorado Mining Industrial Development Board, State Office Building, Denver, Colorado 80202. Visit the mineral display at the museum in Denver for a real treat. They have some gold specimens that will make any prospector's heart skip a beat.

GEORGIA: One of the first gold rushes in this country took place in this state. Many of the 49ers in California learned to gold mine here. All of the gold is located in the northern half of the state. This is one of the most popular spots for East Coast prospectors. The area around the town of Dahlonega in Lumpkin County is the most productive today. Other good areas are in White and Cherokee Counties as well as Warren, Rabun, Hart, Madison, Fulton and other counties in the gold belt. Write to the Georgia Department of Mines, Mining and Geology, 19 Hunter St., S.W., Atlanta, Georgia 30303, for detailed information on the various locations.

IDAHO: Over 10 million ounces of gold has been mined in this state since the 1850s. Boise county has been the most productive, and the Boise Basin the source of most of the gold in the County. The placer deposits of Idaho County have been almost as productive, with the area south of the town of Grangevllle having produced over a million ounces. The Snake river has several areas where you can find gold. It has been worked only off-and-on over the years. Other gold-bearing areas are in Twin Falls, Gem, Ada, Adams, Bingham, Power, Bonneville, Caribou, Cassia, Custer, Elmore, Owyhee, Washington, Valley, Clearwater, Shoshone, Camas and Lemhi Counties. The Idaho Bureau of Mines and Geology, University of Idaho, Moscow, Idaho 83844, can be contacted for more information and maps.

INDIANA: There has been a small amount of gold recovered in the area of south-central Indiana. No records are available as to the total ounces found.

MAINE: Most of the gold found here has been along the Swift River and its feeder streams. There are several other streams that contain gold as well. No record of the production is available, but it has been said that several people are still making a living today working the deposits on the Swift River. There have been some rich pockets also found here. Contact the Maine State Bureau of Mines, Augusta, Maine.

MARYLAND: The area around Great Falls has had several producing gold mines in the past. Not a lot of gold has been recovered, but there are a few placer deposits to be worked.

MICHIGAN: Another state with a small production, about 30,000 ounces is the best guess. The deposits are in Northern Michigan in the area around the town of Ishpeming. Contact the State Department of Mines for maps and details on locations.

MISSOURI: There has been some gold found in several of the rivers here. There is no record of the production and most of the gold found is very fine.

MONTANA: Gold was first discovered here in 1852. Montana has produced over 18 million ounces for prospectors and miners since that first strike. The richest area has been in Alder Gulch, near Virginia City, Madison County, in the southwest portion of the state. Gold is found all along the western border of the state in Lincoln, Sanders, Mineral, Missoula, Ravalli and Beaverhead Counties. The placers in Broadwater County were richer per yard than any other part of the state. Other gold-bearing gravels are found in Deer Lodge, Fergus, Granite, Lewis and Clark, Park, Powell and Silver Bow Counties. For more information, contact the Montana Bureau of Mines and Geology, Montana College of Mineral Science and Technology, Butte, Montana 59701.

NEVADA: Gold is found in every county in Nevada except Lincoln. The principal gold placer deposits are the Charleston, Mountain City, and Tuscarora districts in Elko County, the Lynn District in Eureka County, the Battle Mountain District in Lander County, the Gold Canyon District in Lyon County, the Rawhide District in Mineral County, the Manhattan and Round Mountain Districts in Nye County, the Sawtooth and Spring Valley Districts in Pershing

Huge operation in Nevada, reworking tailings.

County and the Osceola District in White Pine County.

Nevada's lack of water and the rugged remoteness offer a great potential to the prospector seeking good dry washing areas. A letter to Nevada Bureau of Mines, University of Nevada, Reno, Nevada 89507, will get you a list of available publications.

NEW MEXICO: The main gold belt runs through the center of the state from the southwest to a little east of the north-central part of the state, with isolated pockets in other areas. Total production is placed at around the two-and-a-half million mark. Grant County has been the biggest producer, with 500,000 ounces recorded. Other counties that contain placer gold are Colfax, Hidalgo, Lincoln, Otero, Rio Arriba, Sandoval, Santa Fe and Taos. Sierra County is the second largest recorded producer in the state, and the most productive in recent times. Small amounts of gold have also been reported in Mora, Bernalillo, Dona Ana, and San Miguel Counties. Contact the New Mexico State Bureau of Mines, New Mexico Institute of Mining and Technology, Socorro, New Mexico 87801.

NEW HAMPSHIRE: There have been reports of small amounts of gold being found in north-eastern New Hampshire on its border with the state of Maine.

NORTH CAROLINA: A popular area for East Coast prospectors today. Total production for the state is placed at nearly two million ounces. Several large nuggets have been found here. One weighing 17 pounds was found by a young boy in Cabarrus County. The Gold Hill district in Rowan County has been the most productive region, with around 300,000 ounces having been recovered here.

Placer deposits are also found in Burke, Cabarrus, Catawba, Davidson, Franklin, Gaston, Guilford, Lincoln, Mecklenburg, Montgomery, Moore, Nash, Randolph, Stanley, Transylvania, Union, Warren, Yadkin and several other counties. The North Carolina Division of Mineral Resources, State Office Building, Raleigh, North Carolina 27600, will send you a reading list.

OKLAHOMA: The southwestern and southeastern parts of this state have reported placer gold deposits. A couple of small rushes have occurred, but there is no record as to the exact amount recovered. The Spanish are said to have worked mines here with great success during the early days.

OREGON: A very productive state at one time with a total recorded production of around six million ounces. A popular area for gold dredgers today. The most productive regions have been in the southwestern corner of the state and in the area on the eastern border with Idaho. Baker County leads in the amount of gold recovered with over a million-and-a-half ounces being recorded. Gold is also found in Coos, Curry, Crook, Douglas, Grant, Jackson, Josephine, Malheur, Union and Wheeler Counties. Write to Oregon State Department of Geology and Mineral Industries, 1069 State Office Building, Portland, Oregon 97201.

PENNSYLVANIA: There have been reports of small amounts of gold being found in some of the streams here over the years. The only recorded production has been as a byproduct of iron ores.

SOUTH CAROLINA: The gold-bearing areas here are located in the northwestern portion of the state and in lesser amounts along the Georgia border to the southwest. Total production is around 300,000 ounces. The best area today is in Chesterfield County, where present-day prospectors report you can get an ounce or two with a little luck. Other gold bearing areas are found in Kershaw, Lancaster, Chester, York, Union, Cherokee, Spartanburg, Greenville and McCormick. The South Carolina Division of Geology, State Development Board, P. O. Box 927, Columbia, South Carolina 29200, is a good source of information.

SOUTH DAKOTA: This state ranks third in gold production among the states. Most gold has come from one source, the famous Homestake Mine in the Lead district. There is placer gold here in Custer, Lawrence and Pennington Counties. Total production for the state is over 35 million ounces. The South Dakota State Geological Survey, Science Center, University of South Dakota, Vermillion, South Dakota 57069, will provide you with more details on locations.

RHODE ISLAND: There have been several reports of gold being found here, but there is no record of any production. One area is near the town of Glocester and the other is said to be around the town of Foster Center.

TEXAS: Placer gold is said to be found in the panhandle region of Texas, as well as in the central portion of the state. Write to the Texas Bureau of Economic Geology, University of Texas, Austin, Texas 78712.

UTAH: Nearly 20 million ounces of gold has come out of this state. The deposits are found on a hit-or-miss basis all around the state. The following counties have all produced various amounts of gold at one time or another: Beaver, Daggett, Emery, Garfield, Grand, Juab, Kane, Milliard, Piute, Salt Lake, San Juan, Sevier, Tooele, Uintah and Utah. Write to the Utah Geological and Mineral Survey, 103 Civil Engineering Building, University of Utah, Salt Lake City, Utah 84102 for maps and pamphlets.

VERMONT: No records of the amounts of gold mined here are available, but there has been quite a bit found over the years and even today you can still find it. There have been several mines worked here and there was even a minor rush in the late 1800s. One area that most people talk about is around Plymouth Five Corners. Streams like the Rock River, Mad River, White River, West River, Williams, Ottauquechee, Little, Lamoille, Gihon and Missisquoi Rivers all have had gold in certain spots. Even Gold Brook has gold, as do several other rivers. Contact the Vermont Geological Survey, Montpelier, Vermont, for more information.

VIRGINIA: Gold was first reported here in the 1780s, and nearly 200,000 ounces have been recovered, according to reports. The gold belt runs through the central portion of the state. Gold is found in Albemarle, Buckingham, Faquier, Culpepper, Cumberland, Fluvanna, Goochland, Louisa, Orange, Prince William, Spotsylvania and Stafford Counties. This is another potentially good area for East Coast prospectors. Contact the Virginia Department of Conservation and Economic Development, Division of Mineral Resources, Natural Resources Building, Box 3667, Charlottesville, Virginia 22901.

WASHINGTON: Gold is found in many areas of Washington, with the principal region located in the northern part. Production is around three-and-a-half million ounces. Week-end prospectors do very well in some areas today, because deposits have not been worked over like areas in other states. The counties where gold is found are Chelan, Clallam, Ferry, Garfield, Grays Harbor, Kittitas, Lincoln, Okanogan, Pacific, Snohomish, Stevens and Whatcom. For details, write Division of Mines and Geology, Department of Conservation, 335 General Administration Building, Olympia, Washington 82501.

WYOMING: Total production for this state is around 100,000 ounces. Deposits are spread around the state, with Fremont County the most productive region. Gold is also found in Albany, Crook, Johnson, Lincoln, Park, Sheridan, Sublette, Sweetwater and Teton Counties. This is another potentially productive state that has been overlooked by the present-day prospector. Write the Geological Survey of Wyoming, University of Wyoming, Laramie, Wyoming 82070, for details on various areas.

There are other states such as Arkansas and Illinois that have also produced small amounts of gold in the past, but not enough is known about them to give any accurate details.

This list of gold-bearing areas is meant to guide you in the right direction. The most important thing to keep in mind is to look in areas where you know there is gold. I doubt if there is any region in the world that has been overlooked in man's search for gold. The old timers were pretty thorough, and you are better off following their lead. By taking advantage of the information passed down by the other gold seekers who went before you, you can cut down your search time by quite a bit.

Read any books or articles available on the known gold-bearing areas you want to work. Talk to the old timers who live there, visit the closest prospecting, mining, or treasure hunting supply store and find out where his customers have been doing the best.

Someone may wonder, "Why a treasure hunters supply store, I'm a gold miner." There are two reasons: one is the fact that most treasure hunters are gold panners as well, and quite a few go nugget shooting, which is the art of finding gold nuggets with the aid of an electronic metal detector. Some treasure hunters do very well using this method and come up with some outstanding finds. Quite a bit of this has been going on in Australia in recent times and a large number of fair-size nuggets have been found in the land down under.

Another use for the metal detector is finding black sand deposits. I've also heard of some people who have been making some good finds by following leads from lost mine stories. They have taken their pans and dry washers into areas described in the treasure story, and have had some good success working these areas.

In more arid regions, placer gold concentrations may be more erratic, because of cloudburst action. Rain in these areas is infrequent and when it does occur is usually extremely heavy and short-lived, as in the desert where flash floods are a danger. The heaviest rain is always in the mountains, where there is little soil or vegetation to slow the waters as they rush into the canyons and washes. This flooding destroys the slowly-formed deposits and washes them into the deltas, where you will find the gold scattered all through the alluvial fan. With dry placers, you will find the gold at bedrock or near bedrock, but even more often it will be at a higher level on false bedrock.

Gold-bearing gravels in a desert region; note angular shape of gravels. Photo by Larry Winkelman; courtesy of Allied Services

There may be several layers of false bedrock in one deposit, all holding gold in varying amounts. Because of this, the prospector who is drywashing will most often find the paying areas to occur irregularly. This hit-or-miss locating is one of the problems faced by the desert miner.

One of the greatest joys I've had drywashing was hitting a rich pocket in some overlooked area. Many gold-bearing regions have never been drywashed, or have had a very minor amount done. These areas offer a potentially great reward. In searching these areas, keep in mind that the deposits may have little relation to the present topography, as conditions under which the placers were formed may be entirely different from what they are now. That's why newcomers to the field seem to have "beginner's luck;" they just dig any old place and sometimes hit a good lick.

There is really no way to say exactly where the gold will be when dry washing or any other way for that matter, since, again, gold is where you find it. Take that dry washer anywhere gold is known to occur and dig. Here are a number of places to try in the gold-bearing region you are working:

ROAD-CUTS: A bulldozer can do more in an hour than a man with a shovel can in a month. When cutting a road, formations that have never been seen before are exposed. This enables the prospector to see where the different layers of false bedrock are located, when working in areas where there are ancient stream deposits. You can sample all along the different layers to find the most productive spots.

Also, you can easily work under large boulders that are exposed by the cut. Be sure to try the clay deposits under these large boulders, and those under sand and gravel layers. The fine gold traveling with the gravels will work its way down to the clay and become trapped there. Clay deposited by a stream becomes cohesive and plastic and fine gold will be held by it. As the fine gold moves along over the bottom of the stream channel, it works its way under the boulders and becomes embedded in the clay adhering to the bottom of the boulder. Some very rich finds have been made in pockets of clay under large boulders, close to but not on the surface of bedrock. Sometimes they will even cut to true bedrock and below. Road cuts enable you to reach material you never

Old workings. Photo by Larry Winkelman

would see by any other means.

DRY WASHES: One of the best sources of gold from dry washing. I am not referring to the desert only, but to all gold-bearing regions. Any area that has had known production, be it lode or placer, should be checked. Since many regions have creeks and streams that flow only during the winter and spring, and others only in the rainy season, a lot of these areas are overlooked. One of the benefits of dry washing these areas is the ability to work spots that couldn't be reached when the waters are flowing. Most dry washes have a high rate of velocity when they are running, which enables them to carry a great deal of material for short periods. Try to envision how it was in the wash when the water was running and then work the most likely spots just as if you were using a dredge or a sluice box in water. One thing to keep in mind is that the concentration will not be as great as in a stream where the water runs all the time at different velocities, and thus the gold is more apt to be spread unevenly through the material.

CANYON WALLS: Since streams are continually changing the direction of their flow, portions of their deposits are often left isolated. In cutting a deeper channel, a stream leaves benches or terraces along its valley sides. Since these deposits were at one time the stream bed, they offer a potentially good source of gold. In many areas where the present stream bed itself has been worked a great deal, the bench deposits are a richer source of gold than the stream itself. Take your dry washers to the spots along a known gold-bearing stream that has been the most productive in the past and go up to the canyon walls until you find the ancient gravel deposits and work there. Chances are you will do better than you have in working the stream itself. Keep in mind also that these deposits are more than likely the source of the gold found in the stream. One area that is always good is the run-offs down through the bench gravels: a certain amount of reconcentration takes place there.

MINE DUMPS: Some prospectors find good specimens working mine dumps. Many of the early mine operations were not very efficient, and lots of gold was lost. In some areas as much as forty percent of the gold was missed. You might find a nugget here.

TAILING PILES: The same is true as in the mine dumps. Another factor is that the material worked was many times cemented gravels that never had time to fully disintegrate as they traveled through the sluice boxes. I have been to several areas where millions are being spent to work huge piles of tailings. Some of the tailings contain an ounce of gold or more per ton. Now after more than a hundred years, in some areas, these cemented gravels have had time to break down after being exposed to weathering and release the gold they carried. Some very rich finds are being made in working tailings today by dry washing.

GRAVEL DEPOSITS: Any gravel deposit in a known gold-bearing district is a potentially good source of gold. Test both well-rounded and water-worn gravels and the angular gravels of the more arid regions.

MOUNTAIN TOPS: In some areas, ancient stream beds lay exposed high on the top of mountains. Sometime in the past, these

stream beds have been uplifted by great geological forces and are now mountain tops. In some places where they are large enough, they will bear the name "table mountain." It's quite a thrill to climb to the top of a peak and find a bed of water-worn boulders and gravels resting there. Some of these deposits have been worked commercially in the past.

SLOPES: Both the bajadas and pediment deposits in desert areas and even the slopes in mountainous areas containing lode mines are a potential source of placer gold. The desert areas in particular could be a rich source in the future.

DRY LAKES: There have been several dry lakes that have produced gold. At one time these lake beds were fed by streams carrying gold, and if you can figure out the places where the stream entered the bed (and consequently lost its carrying power), you may find some good pay dirt.

BURIED STREAM BEDS: Buried stream placers in desert regions offer one of the most potentially rich deposits remaining. If you should by accident discover one exposed, you would have to search no longer, for you would have found your own El Dorado.

These are some of the best places to dry wash, and you may even have some ideas of your own. Next, we will discuss how to dry wash.

Opposite Page: Old mine building. Many old mines have rich tailings that have never been worked.

Chapter 4
How to Dry Wash

Before setting out on a prospecting trip, check your equipment. Make sure everything is working right before you leave. There is nothing worse than finding that your dry washer is not working correctly after you have spent several hours searching for just the right spot and getting all set up. I've even known people to get out into the field and get all set up and find out they have forgotten the most basic tool of all, their shovel.

Don't forget that many places you will be working will not have any water for panning. I do not recommend dry panning for most people, only a few old timers can really do it well and even they admit that it is extremely slow and recovery is never as good as panning in water.

It is very difficult to save the very fine gold when dry panning. There are several things you can do as far as panning your concentrates are concerned. One method is to save your concentrates until the end of the day and drive to the nearest water and pan them. Before I begin working an area, I learn where the nearest water is located. If you are just out for the day, you may want to wait until you get home to pan out your concentrates. Even when you are out for the whole weekend, you may want to save them. One of the benefits of waiting is that it allows you to spend all your time in the field prospecting. Another popular idea is to bring along a couple of five-gallon drums of water and a wash tub. When you have finished working for the day, you can bring your material back to camp and pan it out that night. Be sure to let the water settle before you recycle it back into the drums, if you decide to move your camp.

Once you have found a spot you want to work, always try to get your dry washer as close to the spot as possible; this way you can shovel the material right onto the trommel screen. You can

Electrostatic concentrator at work

quickly bang up a screen by throwing heavy gravels, which sometimes contains fair-size rocks. With most dry washers, you will be able to throw the gravel you want to work right onto the screen without any sorting. Some of the smaller sampler-type dry washers require classification before they can separate the gold. In the next chapter we will discuss the various types of dry washers available.

If you are not able to set up close enough and have to carry the material in a bucket, be sure you devise some sort of nugget screen to classify the gravels. A nugget screen can be made at home very simply.

To make a nugget screen, all you need is two pieces of wood, 10-1/2" by 2-1/2" by 3/4", and two pieces, 7" by 2-1/2" by 3/4", plus a piece of heavy half-inch screen 10-1/2" by 8-1/2". Nail your wood together to make a box shape open on both the top and bottom. Then nail the screen over the top. You may want to use some molding over the screen around the edge to give it added strength. Whenever you are shoveling gravel into a bucket, place your nugget screen over the top to keep out the larger rocks. Any nuggets larger than the screen will be easy to see, believe me, and you won't have to worry about losing any. The nugget screen will keep you from spending a lot of energy carrying rocks you are just going to throw away.

If you are using a dry washer that has a gasoline motor which is separate from the actual dry washer, make sure that the motor is always on solid ground. The vibration of the motor will cause the legs to work themselves into soft dirt or sand. Some prospectors attach their engines to a piece of plywood to avoid this and to keep dirt from getting into the fan or the engine. Sometimes just a simple rubber mat like you use in a car will do as well. A plywood platform will also keep your motor from dancing around. The platform needs only to be just an inch or two wider than the diameter of the legs. Any thickness of plywood will do.

How long you want to run your dry washer before cleaning out the riffle tray depends on you. Most professionals like to clean it at least every couple of hours. A good rule of thumb is to clean it each time it runs out of gas, normally a little over an hour or so. This doesn't mean you have to pan it out at this time. Keep a small bucket just for holding concentrates and dump your tray into that each time you remove the tray. In areas which contain large amounts of black sand, it is important to frequently clean out the tray, because the heavy black sands will build up behind the riffles and keep the gold from being caught if you go too long. A paint brush is a good tool to use in cleaning out the tray; just be easy with it, or you may flick away some flour gold.

With a bellows-type dry washer, you get a natural vibration from the slapping that occurs as it operates. This is not true with a blower type. Some vibration will increase your recovery with the blower type. This type is very effective and the recovery is excellent until the material begins to build up in the tray.

When build-up occurs, stop shoveling for a while and shake the tray by hand while the motor is running. This takes only a few minutes, and will greatly increase the amount of gold you recover. You need to shake the tray only until the riffles are not overloaded any longer. Do not shake the tray too hard, just enough to allow the heavier gold to work its way down behind the riffles. Sometimes just a series of taps on the side of the tray with your hand is sufficient. Most dry washers are designed with the proper angle to the trommel screen and tray to ensure the greatest recovery. Do not

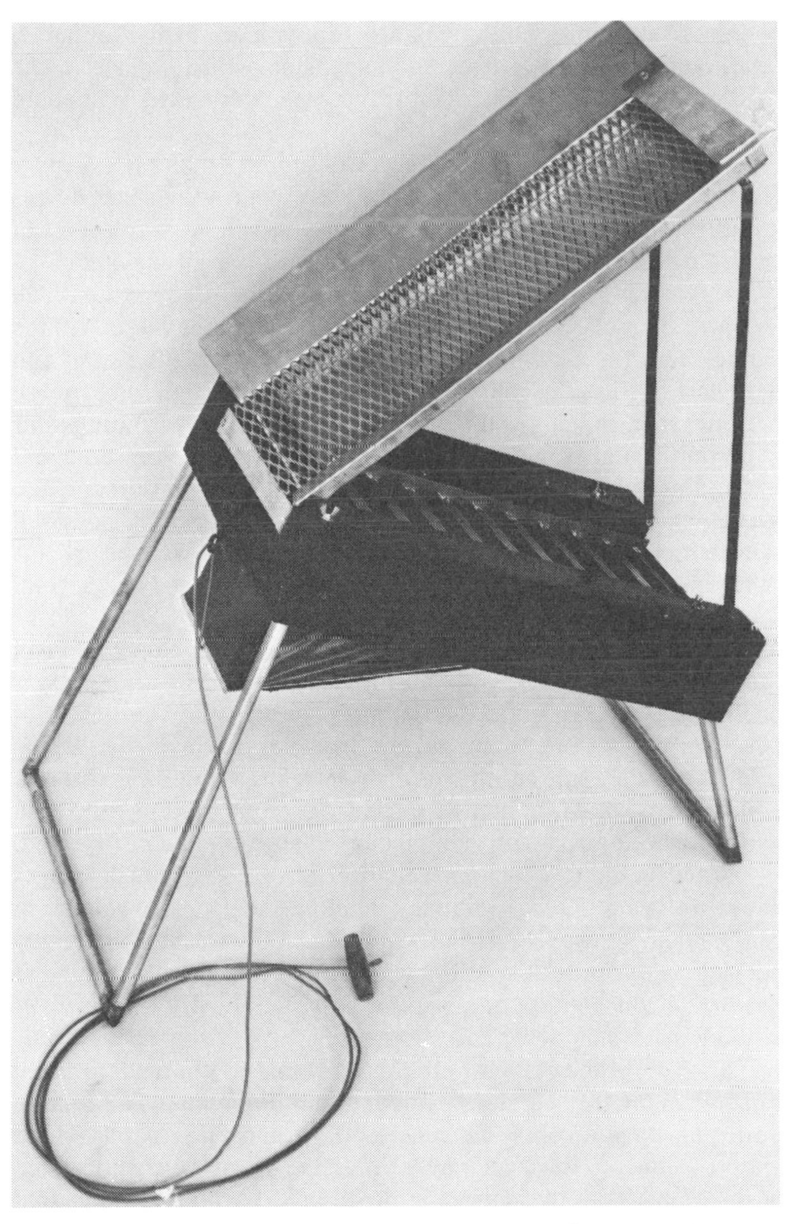

Bellows-type electrostatic concentrator. Courtesy of Keene Engineering

change these angles unless you are sure you are losing too much gold. Some units have a spring suspension which adjusts to the load on the tray. If you allow it to become overloaded, you could cut the amount of gold recovered.

Dry washing is pretty dusty work, so don't make it worse by the way you set up your equipment. Place your dry washer downwind from where you are shoveling, or you will be eating a lot of dust. Be sure that you locate your motor away from the dust.

No matter how arid a region you are working, the ground will begin to be moist after you dig down a few inches or so. This is true of desert areas also. This moistness will cause the sand and gravels as well as the gold to stick together even when run through a dry washer, causing you to lose some fine gold. By running your material through your dry washer more than once, you can avoid losing some of this fine gold. The air blowing through the sand and gravel tends to dry it out and release the trapped particles of gold. Where there is a great deal of moisture, you may want to try another method of drying the material. What you need are several tarps, even the cheap painter's tarps made of clear plastic can be used if you're careful, and don't rip them with your shovel. Spread the tarps out on the ground and throw your gold-bearing gravels on the tarp. Make sure the tarps are placed in the sunniest spot possible, then rake and spread the material out over the tarp as thin as possible and let the sun dry it before running it through your dry washer. This is a particularly good method to use when you are working an area you know has a large amount of fine gold.

When working bench gravels and terrace deposits in places above a flowing stream, you may want to use your dry washer in conjunction with your sluice box. Since most dry washers are highly portable, you can take your dry washer up on the canyon wall and use it as a concentrator or separator only. Let the dry washer classify the material quickly and simply for you.

Just shovel the gravels onto the screen as you normally would; as soon as the tray becomes loaded, dump it into a bucket. Do this until you have a couple of buckets full, then take the classified material down to the river where you have set up your sluice box. Start the material through your sluice box, keeping in mind that this material is concentrated pretty well and a little more caution should be used than if you were just shoveling bank material. Once you have emptied the buckets, return for another load while your sluice box is washing your pay dirt.

Do not mix your material when you are using this method of prospecting, or any other time for that matter; this is to keep you from someday wanting to shoot yourself. One of the most frustrating things that can happen to you is to be panning out concentrates that you have gathered from several areas and find them to be very rich for just one pan or two; that means one of the spots was hot and you can't be sure without going back and checking each area all over again. Do not throw any bank gravels in the sluice box along with the bench gravels you are running.

One last note. Don't be in a hurry when you're dry washing. The old saying, "haste makes waste" really holds true with dry washing. It is not the same as when you are using water to wash for gold. Always remember that the gold's density to the mass of other material is greater in water than when you are dry washing, and you have to work a little more carefully or you will lose a lot of fine gold. If you follow these rules, you will be coming home with more gold than ever before.

Chapter 5
Equipment

Over the years, hundreds of devices have been developed for the recovery of gold from placer deposits by dry concentration, most of which have been quickly discarded as impractical or inefficient. Most people are not aware that the great inventor, Thomas A. Edison, about 1897, designed a dry process machine for saving gold from a placer deposit in New Mexico. It was not reported how effective it was .

One of the earliest methods of dry separation, winnowing, was brought to this country from Mexico. It is a rather crude method of concentrating the placer gravel by exposing it to wind, while rolling and tossing it in a blanket. The finer particles are blown away and the coarser material is picked out by hand, while the fine gold is trapped in the fibers of the blanket.

There are three basic types of dry concentrators using air as means of separation. One is the type that projects the material to be concentrated through air by force other than an air blast, the particles being classified by their momentum. The second is machines using a continuous blast of air, and the third is machines using intermittent pulsations of air, where the heavier minerals settle and the lighter material comes to the top and is removed by gravity and air. We will not concern ourselves with the first, because it is not commonly used in the field today. The second method is a blower-type dry washer and the third is a bellows-type dry washer.

You can build your own dry washer, or buy one ready to go. Of the do-it-yourself types, the most common is the Mexican dry washer, or air jig. Along with the manufactured blower-type devices, these are especially satisfactory for small-scale placer mining.

Even though there are a number of designs and styles of bellows-type dry washers, the same principle is involved in the con-

Winnowing for gold

struction of all of them. The dry washer operates on the same principle as the jig, and consists essentially of a screen, hopper, riffle tray, and bellows, all mounted on a wooden frame. The bellows is normally made of canvas. The air enters the bellows at the bottom through a hole covered with a leather flap and is forced up through the riffle tray, blowing up and removing the lighter material, while the heavier minerals and gold are trapped behind the riffles. The larger material is moved by gravity and vibration. Most home-made

The Gold King dry washers

dry washers are constructed of wood and weigh (depending on the size of the unit), between thirty and forty pounds without a motor. The riffle tray will be around twelve inches by twenty-four inches; the rest of the machine will be scaled to it.

Some of the best factory-made bellows-type dry washers come from Keene Engineering in Northridge, California, which is the largest manufacturer of gold mining equipment in the world. One of their most-popular units is their Model DW212V, which weighs

The author using Keene's lightweight, hand-cranked dry washer

approximately forty pounds, and can be operated with either a hand crank or a twelve-volt electric motor. It has a Marlex plastic base, which is strong and lightweight, as well as an adjustable-flow control gate, providing an even flow of material over the riffles.

The best addition is the twelve-volt electrical option. By adding the small electric motor, you can eliminate hand cranking. The motor works on any standard twelve-volt battery. For most operations, a regular car battery is recommended. If you are packing into a remote area with the dry washer, a small motorcycle battery will give you around four hours of operation. Even with the motor, the unit is still portable and is easily carried on a standard backpack frame.

If toting a battery is not your style, you may want to settle for their smallest unit, Model DW2, which is strictly hand-cranked and weighs only thirty-two pounds.

Keene Engineering has made several improvements from their original mini dry washer. They have installed a thermos plug in the end of the box, so that you can easily recover the material that has fallen through the fabric on the riffle board, instead of picking up the whole unit and turning it upside down to dump the concentrates that filter through. It has a removable riffle tray that can be dumped into a bucket or pan—a much-needed improvement that avoids a mess and saves fine material.

Their original unit worked by means of a pull-rope on the bellows; pulling that rope handle sure got tiring, and could raise blisters on your hands. Now, with a hand crank, it is really easier to operate—and faster as well.

The Keene Electrostatic Concentrator is one of the best machines available to the small-scale placer miner. I got one years ago when they were first introduced and still have it. Mine has been trouble-free and has recovered a lot of gold. The electrostatic process is not new, but you may not be familiar with it.

Gold is non-magnetic, but it does have an affinity for an electrical charge. The Keene Electrostatic Concentrator uses a high-static air fan to force air into the Marlex plastic base, where it gets a charge from the plastic. From there the air moves under pressure

Home-built unit at work. Photo by Larry Winkelman

through a special cloth under the riffles, which creates an even-greater charge. This electrostatic charge in turn passes on to the gold particles, causing them to be attracted to the cloth and stick to it. (An example of an electrostatic charge can be seen when pieces of paper are attracted to a comb after the comb has passed through your hair.) The paper is not affected by a magnet, but still will be attracted to the comb by an electrostatic charge.

The electrostatic principle keeps you from losing a lot of the fine gold that some conventional dry washers miss. The steel frame is designed so that overflow tailings drop directly on the legs to provide a stability that is equal to machines many times its weight.

The power for the Keene Electrostatic Concentrator system is generated by a high-static air pump coupled directly to a 3-1/2 horsepower Briggs and Stratton engine. Both engine and pump are made of aluminum. The mounting system permits heat from the engine to transfer directly to the air pump, pre-heating the air and making it possible to process damp sand that cannot be put through some other units. For added measure, I run the damp sands through more than once to make sure that they are good and dry. The total weight of the Electrostatic Concentrator, including the motor, is 78 pounds. One man by himself can carry the whole unit into almost any area. I have seen miners rig up several of these units in a row and run a large amount of material in a day.

One of the more recent innovations in dry washing is the air-driven vibrator. In these machines, vibration is provided by air passing through a fan with offset weight. The combination of the vibration and air flow greatly increases recovery of fine gold.

The Gold King is another line of outstanding dry washers. They make several models, designed to fit any prospector's needs. The all-metal construction is built to withstand the abuse dry washers receive out in the field. Other features are blind riffles, which create dead-air space to trap the microscopic gold, and an internal vibrator that allows you to decrease the normal trommel and riffle angles for maximum efficiency and greater recovery.

Their Dust Devil family includes The King, The Queen and Junior units. The largest unit, The King, has a twelve-by-thirty-six-inch riffle box and is powered by a Briggs and Stratton 3-1/2 horsepower engine. Total weight, including blower, is around eighty-five pounds. The others feature a lightweight, high-volume blower. The

Dust Devil Junior weighs thirty-five pounds and can be used in conjunction with their vacuum system, since the same common blower can operate both machines.

A nice feature is the tray in the riffle box, which is designed to fit so tightly that very little material gets by it into the bottom of the box. Gold King recommends removing it only at the final clean-up. I carry a small paint brush to sweep the fines into my pan. Most stores carry the Gold King line, and they are worth looking at when you are searching for a new unit.

There are many, many variations of the home-made type of unit. I've seen big ones, little ones, fat ones and skinny ones. I've also found a lot of home-made units left behind by unhappy prospectors. Make sure the machine you choose is efficient, highly portable, and very durable. Dry washers take more of a beating than any other type of gold-saving device and must be built to last.

There are several small manufacturers of dry washers who more-or-less work out of a home workshop and make a good unit. Most of these dry washers are made of wood.

TOOLS

Some of the other tools you will need on your dry washing outing are a pan, shovel, tweezers, magnifying glass, knife, brush, bucket, compass and a prospector's hammer. If you already have a metal detector, you may want to take it with you. It is not really required, but sometimes can be an aid locating black sand deposits. Most detectors can do this; more information on the various instruments can be obtained from your local dealer.

THE PAN

The basic item used by both the beginner and the old pro is still the pan. During the rush to the Mother Lode, almost anything that was available was used as a gold pan. Many old-timers cooked and panned out of the same skillet!

The Mexicans were probably the first to bring a vessel to the gold fields for the express purpose of separating precious metals from the sand and gravels of stream beds. The Mexican pan was called a *batea,* and it was carved from wood. It was fifteen to sixteen inches wide, six to eight inches deep, and fairly heavy.

An interesting story regarding the *batea* is told about James

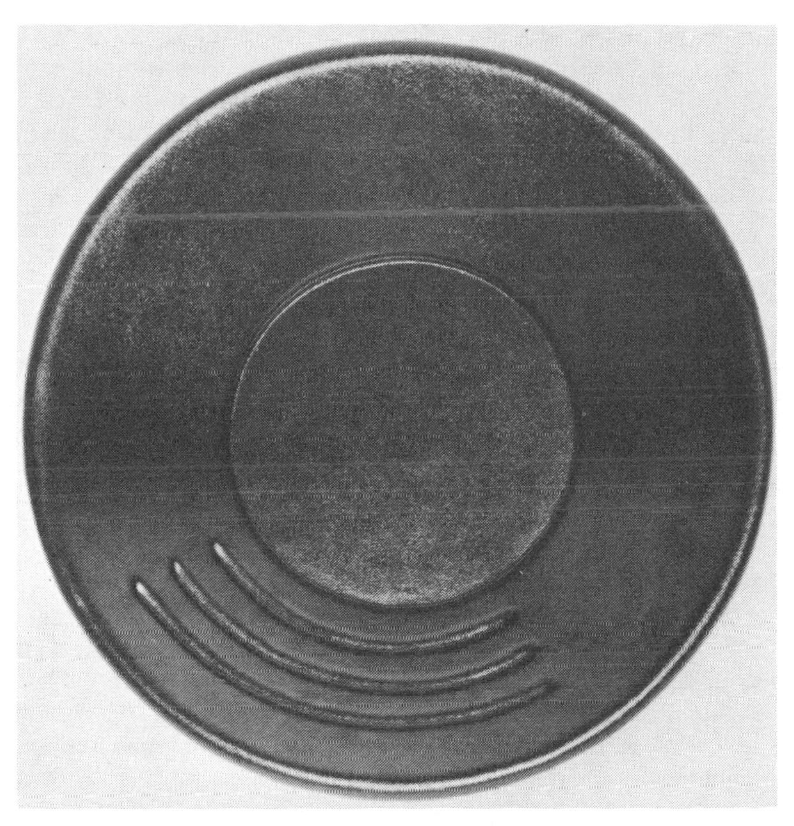

The plastic pan; note the riffles built into the side of the pan

Marshall, the discoverer of gold at Sutter's Mill. This discovery started the California Gold Rush, of course. It seems that one of the workers at the mill, who was from Mexico, had told Marshall that if he had a *batea,* they could recover gold from the gravels of the river for themselves. Marshall thought that the *batea* was some sort of complicated device, so he put the man off. Whether one *batea* would have helped him, we'll never know, but Marshall saw little of the millions in gold that resulted from his discovery. He died a penniless and bitter man.

Today's pans are made of steel, copper, or plastic. For the beginner, I would recommend the plastic pan for several reasons. One reason is that it is molded with a set of cheater riffles in the pan, that keep a new panner from losing a lot of gold by acting as traps for the gold. The pans are black, so the color contrast with the gold makes the metal easier to see when separating it from the concentrates. Steel pans, on the other hand, must be burned black when they are new. Another good thing about the plastic pan is that it will not rust or become corroded if wet black sand is left in it. The pros use copper pans when using mercury to recover fine gold.

Three sizes of pans are made today. The small sampling pan is about six to eight inches wide; the medium pan (which most beginners find easiest to handle), is twelve to fourteen inches wide. Stay away from the big one in the beginning, because it is awkward to carry around and collects too much material for easy panning. These are sixteen to eighteen inches wide. It also gets pretty darn heavy! Most hardware stores carry the steel pans. The plastic and copper pans can be purchased at prospectors' supply stores, rockhound shops and some sporting-goods stores.

WHITE ADHESIVE TAPE

A roll of white adhesive is a must. No, it's not for doctoring; it's for marking your ore samples. Whenever you locate a vein that you feel might contain values and you think they should be assayed, use the tape to mark the location. God only knows how many pieces of rich ore I've seen that came from some unknown place; no one knows where they came from.

Write the date and the exact location on a piece of tape and stick it on each piece of ore you bring in from the field as a sample for assaying

THE SHOVEL

You'll need a good shovel to dig up the dirt to put into your pan. Take your choice: long-handle or short. The long-handled shovel is easier on the back; the short-handled one is easier to carry. You also should have a small hand garden trowel for hard-to-get-at spots.

TWEEZERS

You will need a pair of pointed jeweler's tweezers to pick the tiny flecks of gold from your pan.

SMALL BOTTLE

Find a small bottle or vial of clear glass or plastic (with a reclosable cap) to hold your colors.

MAGNIFYING GLASS

Any small magnifying glass that you can carry in your pocket will do. For home study, the magnifying glass attached to a flashlight is the best.

KNIFE

A plain hunting knife will do. Use this for digging, prying and scraping.

BRUSH

A small paint brush can be used for cleaning out cracks and dusting crevices which may be hiding gold, hopefully.

BUCKET

Any regular household plastic bucket will suffice. The size will depend on how much you will want to carry.

A large, motor-driven dry washer from the 1930s.
Courtesy of the Arizona Bureau of Mines

COMPASS

Pick up a good compass before you go; it may save your life. Hundreds of stories are told of people losing their lives wandering aimlessly only a few miles from a main highway. Select a prominent landmark and note its direction from your campsite. Note also in which direction the nearest main road lies.

A PROSPECTOR'S HAMMER

You may want to invest in a prospector's hammer, which is used to split rocks to see if they contain any valuable minerals. You can also use the hammer end to pry. It is pointed on one end and flat-headed on the other. Buy a good one, otherwise the point on the cheaper ones will flatten out after very little use.

These are the basic tools you should have to start prospecting. As you continue in your golden quest, you will want to acquire more sophisticated equipment, such as a sluice box, dredge or metal detector.

Chapter 6
How To Build A Dry Washer

This chapter was written by Carl Fischer, and is courtesy of *Treasure* Magazine. This is one of the best-designed home-built units I've seen and the plans are explained so that anyone who is handy with tools can do it himself.

Only a rank amateur would expect to strike a rich bonanza today in the running stream bed of a well-known gold area. This does not mean that the Mother Lode Country has given up all of its gold. Far from it. There's more gold there now than was taken out during the Gold Rush Days! But those wet deposits have been worked over by so many people so many times that your pickings are bound to be lean if you plan to sift over the same gravel beds that have gone through countless sluice boxes in the last hundred years!

You *could* find that rich bonanza if you prospect the banks of these same streams fifteen, twenty or even a hundred feet above their present water levels. If you look for gravel deposits on overhanging ledges and in the bends of the stream which were laid down hundreds or thousands of years ago, you'd be prospecting in virgin territory. These deposits are not likely to be continuous or to contain extensive material, but they are easy to spot if you look for layered and stratified gravel and sand. Many deposits are small and contain only fifteen or twenty yards of material, but they are apt to be extremely rich in gold, because the rivers were loaded with it a thousand years ago. Unless the ground cover is unusually heavy, a pair of binoculars or a spotting scope will help you locate these higher deposits, especially from the opposite bank.

Besides being extremely rich in gold, these areas all have one thing in common—they are bone-dry most of the year. In fact, you may need a pick to break up the deposits so that you can get them into your recovery equipment. Because these deposits are dry, a modern portable dry washer is exactly the right instrument to use in these upper levels. If a flat area large enough to accommodate your machine is not available at the site, scout the cliff ten or fifteen feet below for a suitable one. Several ends of corrugated plastic roofing tied into a trough make an excellent chute to slide the material down to a lower level where there's sufficient working room.

If you're thinking what I think you're thinking—that a dry washer will reclaim large gold, but is not efficient in recovering fine gold—perhaps you haven't heard what happens when a blast of air is forced through a nylon screen in contact with cotton cloth. This combination sets up a charge of static electricity in the cloth which acts like a magnet to hold the fine gold. Under a magnifying glass, these fine gold particles can be observed actually standing on end as they cling to the cloth above the riffle. A properly built dry washer can reclaim fine gold down to 200 mesh. It's the nylon screen in the modern dry washer, plus the proper inclination of the tray, that makes it an excellent recovery instrument.

If you're handy with tools, you can build a dry washer of this type in your spare time. With the exception of the motor and jack

Close-up of a home-made unit. Photo by Larry Winkelman. Courtesy Allied Servics

shaft, the cost of materials is less than $60. The unit I use is portable even when it is fully assembled, and it knocks down for easy storage in camper or car trunk. Dimensions may be changed to process larger loads, but the specifications given here will handle one-and-a-half-tons per hour, on the average. The machine, less the motor, weighs under 35 pounds. A bill of materials is listed under the picture of each component part. The unit is built of 1" x 2" construction-type lumber, unless otherwise noted. Wood screws used are 1-1/2" #8; bolts are all 1/4," machine or carriage types, as specified. Wing nuts are used on all parts so they can be disassembled for storage or transportation.

THE RIFFLE TRAY

We start with the tray because this is the principal component; its size sets the pattern for the rest of the machine. Cut two pieces of 1" x 2" lumber exactly 25-3/4" long and another piece 14-3/8" long. Lay out the pieces on ends and fasten the shorter one flush with the two longer pieces using two screws at each joint. Use white glue between joints and drill all screw holes with a 3/32" bit to prevent splitting the wood. (Rub your screws over a wet piece of hand soap before driving—it makes the job a lot easier.) The tray shall be exactly 14-3/8" wide. Lay the finished piece on a flat surface to assure alignment. The short 1" x 2" piece marks the head of the tray. Measure 24" back from the head of tray and rabbet out a 1" groove, 1/4" deep, and insct a 1" piece of 1/4" plywood here.

Turn the tray over and tack a piece of Indian head cloth (either red or white) to cover an area 24" x 14-3/8" on the bottom. The cloth should lie smoothly, but it should not be stretched. Over this cloth tack a piece of nylon screen, the same size as the Indian head cloth. Roll it on smoothly, but do not stretch it. Cut five batts from the 1" lattice and tack these to the back of the tray, 4" apart. The top batt starts 4" down from the head of the tray. Use one-inch #18 brad nails for tacking the batts. Cut strips of 1/4" plywood, 3/4" wide, to flush in the gaps between the batts so the surface is smooth. Fit these pieces as closely as possible, then cut pieces of 3/4" heavy weather-stripping to line the lengths and ends of the bottom of the tray.

Lay the tray face-up on your work bench and cut five pieces of

1/2" quarter-round molding 13" long. Tack these to the batts, using 3/4" #18 brads. The straight side of the quarter-round shall face upward on the tray, and the curved bottom part shall flush with the bottom of the batt. The reason for this arrangement is evident in the silhouette picture of the tray; this area is a 1/2" dead air space in front of the riffle, which gives the heavier deposits a place to stick. This is where you will find the gold in your tray! Put the tray to one side now; we'll install the snap catches when we are finished with the bellows box.

THE BELLOWS BOX

Cut 1" x 2" lumber to specifications shown under the picture of the bellows box. Lay out the two long pieces on the work bench and cap them with the 14-1/2" piece, using two wood screws at each joint. Measure 1-1/4" from the foot of this box; at this point fasten one of the 13" members. This piece will form the foot of the box. The other 13" piece forms the top of the bellows and is fitted in 12" from the head of the box, measured from the outside of the cross member. The bellows box shall be 24" long, outside to outside, exactly the same dimensions as the tray.

Now cut a piece of 3/8" plywood 24" x 13." This will be the movable part of the bellows. About 8" from the bottom end of the plywood and on a centerline, cut a 2-1/2" round hole. Use a circular saw bit that fits your electric drill for this operation. Cut partway through the piece from one side; then turn the board and finish the cut from the other side, in order to eliminate rough edges.

To cut out your bellows, lay out a piece of heavy blue denim, 78" x 12," with the selvage edge towards you. Mark the centerline of this material with a white or red pencil, so it will show up on the dark blue cloth. Measure two points on each side of this centerline 6-1/2." This 13" part will make the head of the bellows. Next, mark off two points from both ends of the denim, 7" in from each end and 3" above the selvage. Using a yardstick, draw lines from these 7" marks to the two 6-1/2" points you measured earlier at the center. Cut along the lines and you have the material which will be the bellows for your box.

SPECIFICATIONS:

Screen and canvas trommel
Specifications:
2 pieces—1"x2"x39"
1/2'' hardware mesh 14" x 36"
1 piece—13" x 13 3/4" 1/4" plywood
1 piece 1" x 2" x 13 3/4"
5 -1/2" 3/4" molding 1/4" thick
1 piece—24" x 30" 10 oz. duck

Specifications: Plywood triangle and wing nuts allow quick assembly of unit. Motor drive fastens to bolt holes in base.

A-frame
2 pieces—1"x2" x 57 3/4"
1 piece—1"x2" x 43 1/2"
2—1 1/2" machine bolts and nuts
2 pieces—1" x 2" x 54"
2 pieces—6" x 10" 3/8" plywood *(for triangles)*
2—2 1/2" Carriage bolts and wing nuts

Hopper
Specifications:
2 right triangles 9" x 10 3/4" x 14 1/2" of 3/8" plywood
1 piece—14"x14 1/2" 1/4" plywood
1 piece—3 1/2 x 4" pressed wood *(for gate)*
1 piece—9" x 14" 1/4" plywood
1 piece—3/4" x 10" 3/4" AL.
2—2 1/2" Carriage bolts, wing nuts

Top of Tray
List of materials:
2 pieces—1" x 2" x 25 3/4"
1 piece—14 3/8" x 1 1/4" (1/4" plywood)
1 piece—Indian head cloth, 15 1/2" x 25"
5 pieces—1 1/4" x 14 3/8" *(1/4" lattice)*
2—snap catches
1 piece—1"x2"x 14 3/8"
5 pieces—1/2'' quarter round 13" long
1 piece—Nylon screen 15 1/2"x25"
8 ft. 3/4" Heavy weather stripping

Upper part of bellows box.
Piano hinge occupies lower position in unit.
Specifications:
2 pieces—1" x 2" x 36"
2 pieces—1"x2"x13"
1 piece—1" x 2" x 18 1/2" oak (tongue)
7 lf. heavy blue denim, 12" wide
Box of 3/8" small-head tacks and a 13" piano hinge, 1/2" wide
4—2" Machine bolts and nuts 1/4"
4—2" Carriage bolts and wing nuts
1 piece—1"x2"x 14 1/2"
1 piece—13" x 24" 3/8" plywood
1 piece—1" x 2" x 10 1/2" oak (coupler)
5 ft. 3/4" molding, 3/4" thick
1 piece—Heavy rubber 8 1/2 x 4 1/2
2—Snap latches
1—2 1/2" Machine bolts and nuts

Mark the center of the top of the bellows box—this is the second cross member. Match this mark with the centerline of the denim, placing the selvage edge about at the center of the 1" x 2" cross piece. Fasten the selvage edge to the inside of the bellows frame, using small-head 3/8" tacks. Start at the centerline at the head and work back, down both sides to the foot of the box. When you get to the foot, there should be a lap of about 2." It is not necessary to tack this closely—a tack about every 3" is good. These edges will be reinforced with molding later.

We are ready now to install a 13" piano hinge to the foot of the 3/8" plywood, which is to serve as the movable part of the bellows. The foot is the part nearest the round air-hole. Saw this hinge with your hacksaw so that it fits exactly between the longitudinals of the bellows box. It will be necessary to use short 3/8" wood screws for this operation. Be sure to fit the hinge as shown in the picture so the unit will swing properly. Fasten the hinge to the plywood first, then fasten it to the inside of the foot of the bellows box, over the blue denim.

Turn the box over on your work bench and line up the centerline of the denim with the center of the head of the plywood. Fold under about 3/4" of material before tacking, and work both ways from the centerline at the top. At the upper corners, take a 1" gather in the material to get a better fit. It is not necessary to gather material at the lower corners. Tack the denim to the *under* side of the plywood with 3/8" small-head tacks, spaced not over 1" apart. Do not

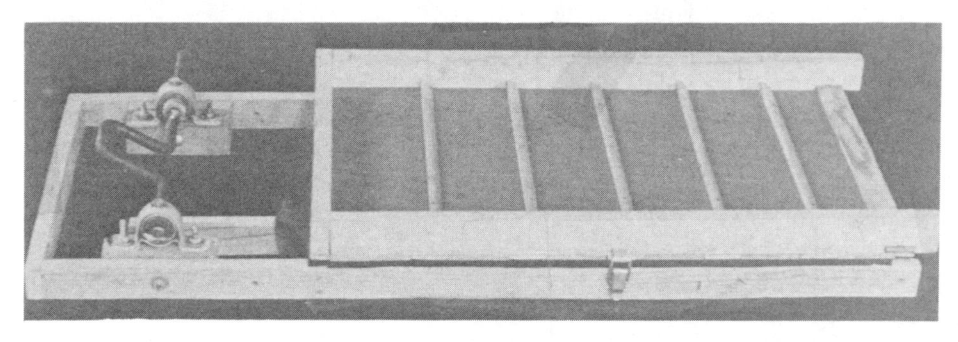

Tray locked to bellows box. Note that straight side of quarter-round faces top of tray. Holes in bellows box fasten unit to A-frames with 1/4" carriage bolts and wing nuts.

use longer tacks because they will protrude into the inner side of bellows and may damage the cloth. When finished, turn the box over and tack the molding to cover the edge of cloth on the upper part of the box. It is not necessary to reinforce the hinge area.

THE HOPPER

My machine has a hopper with a special gate to control the flow of material into the tray. At no time should the tray be allowed to run above the 2" sides; otherwise, you might lose some of the heavier concentrates. Square-cut a piece of 3/8" plywood, 10 3/4" x 9," and cut this rectangle on the diagonal to obtain two right triangles which will form the sides of the hopper.

Fasten the smaller piece of 1/4" plywood shown in the specifications to the 9" side of the triangle with glue and five of the 3/4" brad nails. Notch out a two-inch square hole in the center of the larger piece of 1/4", as shown in the picture. Next, cut four pieces of 1/4" plywood 3" long—2 pieces 1-1/4" wide and 2 pieces 1" wide—and glue these together, flush on one side. Then cut one rectangle from 1/8" pressed wood, 3-1/2" x 4", for the sliding gate. Lay this rectangle over the 2" hole, center it, and fasten the guides to each side of the plywood with glue and brads, allowing sufficient space for the gate to slide easily. This need not be a tight fit.

Drill a 3/16" hole 3/4" from one end of an aluminum strap and a 5/16" hole six inches from that end. Pass a 3/4" long, 3/16" threaded screw through the aluminum strap. Mount one nut on the screw, to be set later. Drill a 3/16" hole through the top center of pressed wood and pass the screw through this. Mount an end screw flush with the end of the screw and tighten the inner nut to hold the gate tightly. Now glue together two pieces of 1/4" plywood, 3" long and 1" wide, and mount this combination on the box in position to serve as a fulcrum for the lever. Drill a 1/4" hole through the mounted 1/2" piece and the plywood side of hopper and mount a 1/4" stove bolt, 1-1/4" long, through the hole from the inside of hopper. Place a washer on the bolt, pass the bolt through the lever, place another washer on the bolt and attach a wing nut. Now you can control the gate opening with pressure on the wing nut and can set it at any desired opening.

To make the stop for the backward swing of the hopper, cut a 4" square of 1/4" plywood and fasten it with glue and a few brads to the center bottom of the 9" back of the hopper. Let the edge of the square protrude about 1" below the bottom of the hopper, but not low enough to drag across the cloth of the tray. This stop should touch the cross member at the head of the bellows. It also serves to keep the gate from sliding out of the guides when the wing nut is loosened.

THE DRIVE ROD

If you have an acetylene torch, you can bend the 25" steel rod to proper specifications; otherwise, take it to a welder and let him bend it for you. A propane torch will not supply enough heat to bend this piece properly. If you do it, lay the rod on a flat cement floor and keep it as straight as possible, because any unevenness will cause the shaft to wobble and this may wreck your machine. The drive rod should be bent to the following specifications:

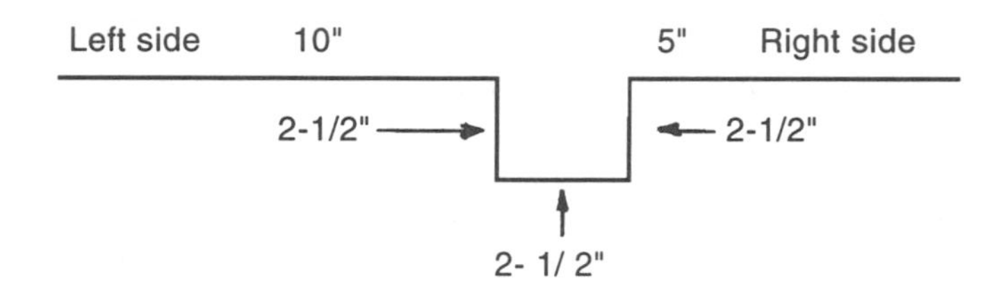

The drawing shows the center line of the rod. Since 1/4" of rod is used in each of the bends, the finished drive shaft should be 17-1/2" long, overall. *Note:* Operator's normal position is behind the screen; therefore, right side and left side refer to right and left of the operator's position.

Smooth out the rod with emery cloth at points where the bearings will ride and push on the right side of shaft. Smooth out left side of shaft and push on one half-inch stop, a bearing, and the over half-inch stop. It is not necessary to secure the stops at this stage. Mount blocks of 1" x 2" lumber inside the bellows box (above the bellows) and mark drill stations for bolts to hold bearings in position. Be sure that at least 2-1/2" of shaft extends beyond the outside longitudinal on the left side of the bellows box. Drill 1/4" holes through lumber and mount four 2-1/2" carriage bolts to hold bearings in position. Slots in the mounting will allow positioning of the bearings so that the shaft turns freely without binding in any position. After bearings are fastened to longitudinals, fasten stops on both sides of the left bearing.

If the drive shaft is shorter than needed for this machine, use 1-1/2" blocks on the sides to properly position the shaft. If your shaft is entirely free of wobble, short blocks will suffice; but if there is the slightest wobble, run the reinforcing lumber all the way from the headboard to the first cross member. This shaft will turn at about 200 rpm and should be fairly straight to prevent excessive vibration.

Prospectors used to build their own dry washers—and many still do. Photo by Nell Murbarger

The hardwood coupler is made from a 10-1/2" length of 1" x 2" oak. Drill two 1/2" holes through the 2" side of this piece, one which is 1/2" (center) from the end and the other 3" (center) from the same end. These will be the bolt holes to keep the coupler together. Now lay the piece flat and drill a 1/2" hole through the center, 1" from the same end from which earlier measurements were made. This will be the hole for the shaft. Draw a ruled line through the center of the 1/2" hole, 4" back from the end and saw out this 4" piece of oak. This is your coupler. Drill another hole through the other end of the oak, 1" from that end. This will take the bolt which holds the coupler to the tongue. Then pass the shaft into the 1/2" hole of the coupler and fasten with two 2" machine bolts with washers on each side.

The other 18-1/2" piece of oak is for the tongue. It is mounted upright on the bottom of the bellows box. Draw a ruled line 6-1/2" from one end down the center of the oak and saw this piece out—it will not be used. Now drill two 1/4" holes through the narrow part of the tongue, 1-1/4" from the end and the other 4-1/2" from that end. These holes will take the machine bolts that fasten the tongue to the bottom of the box. Three-fourths of an inch from the other end of the tongue, drill a 1/4" hole which will take the bolt holding the tongue and coupler together.

With the coupler temporarily attached to the shaft, position the tongue on underside of the box (plywood part) so it extends 8-3/4" beyond the edge of the bellows box. Pass a bolt through the coupler to position this piece properly and mark the plywood (through drill holes in tongue) for positions of the two bolts that mount the tongue. Pass two 1/4" machine bolts through the plywood and tongue and fasten with flat washers and nuts. Use a 2-1/2" machine bolt with three washers—note one washer between the coupler and tongue—and fasten with nut. This joint should not be drawn up tight. After all adjustments have been made and unit is ready to put into operation, saw off any protruding ends of bolts with a hacksaw and hammer down on edge threads of bolt to seal the nut. This should be done on three bolts—two in the coupler and the one holding the coupler and tongue.

Mount the 12" pulley on the left side of the shaft, allowing sufficient clearance, and set this member aside while building the A-frames which hold the unit in position. Cut two pieces of 3/8" plywood according to the following specifications:

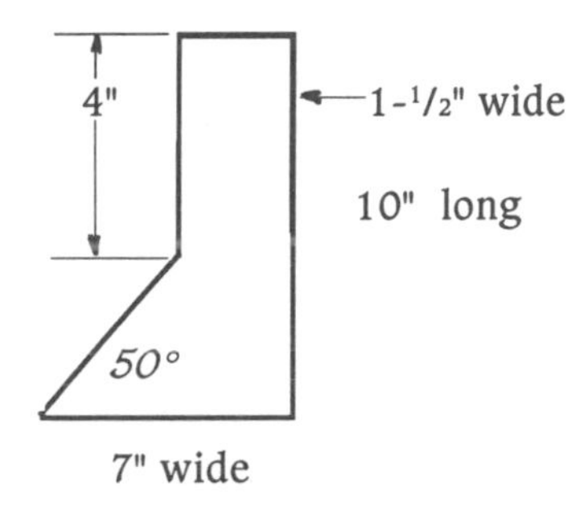

Mount these two pieces on each of the 54" standards. Fasten with glue and three 1" wood screws so that top and flat sides are flush with the 1" x 2" lumber. Position the 57-3/4" part of the A-frame so that it is flush with the slant of the plywood and draw a ruled line down one end to fit. Cut on a 40° angle (complement of the 50°). Cut in miter box to be sure. Drill a 1/4" hole through the standard and plywood and fasten with 1-1/2" machine bolt, washer and nut. Do not glue this joint because it is set for easy disassembling.

Position the 43 1/2" piece across the foot of these two standards and drill 1/4" holes through both, 1" from the ends. Pass a 2" carriage bolt through the holes and fasten with washer and wing nuts. This is the *left* side of the A-frame. When assembling the unit, the plywood triangle shall be on the *inside* of the standard. Follow the same procedure for the standard on the right side, but this side does not require a base runner.

Now lay the two A-frames on your work bench and drill 1/4" holes through their centers at the following positions. All measurements are from bottom of standards. Drill a hole in each of the short (54") standards 12" from the bottom. Then drill one hole in each of the long standards (57-3/4") 33" from the bottom and one hole 45-1/2" from the bottom. These holes will take the bolts which hold the bellows box and the hopper to the A-frame. All demountable parts in the unit are fastened with wing nuts, so they can be assembled and disassembled with great ease. Drill out all of these holes with a 5/16" bit so the bolts can be removed more easily.

Cap the two 39" pieces of 1" x 2" lumber that make up the screen with the 13-3/4" piece. Glue and fasten with two 1-1/2" #8 wood screws at each joint. Lay this on a flat surface and keep the longitudinals parallel—tack a strip of 1/4" plywood to the end to temporarily hold it in place. Now, cut a piece of 1/2" hardware mesh 14" x 36" and tack this to the frame with staples 6" apart, then cut a piece of 1/4" plywood 13" x 13-3/4" and nail this to the lower end of the underside of the frame, overlapping the bottom part of the screen.

You will need a 24" x 30" piece of 10-ounce cotton duck to serve as a trommel under the screen to guide concentrates into the hopper. It is to extend down the screen 24" from the top. Gather it in the center and tack the top of this material over the screen, letting the bottom flare out. Now, take the edges and tack them to the sides over the 1/2" screen. The bottom should balloon into the hopper when unit is in operation.

Using 3/4" molding, 1/4" thick, cover the duck and screen ends on the top and sides of the frame down to the plywood, fastening this with one-inch #18 brads. Cut off any protruding ends of wire mesh that might cut or scratch. Next, drill 1/4" holes to hold the screen to the A-frames. Measure from the top of the A-frame 1-1/4" on each side and drill. Measure from the outside top of the screen down 9" and drill holes in each side of the screen. Enlarge these holes with a 5/16" bit. Mount the screen to the A-frames with 2-1/2" carriage bolts, flat washers and wing nuts.

Cut a piece of rubber inner tube 6" x 9" and tack it over the air hole in the bellows box, one tack at each corner of rubber. Fit the tray onto the bellows box. If there is a visible space between the two, add a strip of 1/4" weather-strip to the corresponding part of bellows box. When units are in perfect alignment, fasten the snap latches to both—the larger part with the snap latch mounts on the bellows box and the catch on the tray. Mount the screen between A-frames and tighten the wing nuts. Drill 1/4" holes in the side members of the bellows box, 3/4" from the end of the upright. Mount the bellows box and hopper in holes previously drilled in A-frames, drilling corresponding holes in the box and hopper where needed. Slant of the tray should be between 35° and 40°.

If you plan to use a two-horsepower gasoline engine to drive your dry washer, you will need to reduce rpm. With a 1" pulley on the motor shaft, transfer power to a 4" pulley mounted on one end of a jack shaft, to a 3" pulley on the other end. Throttle back engine speed to about 3200 rpm, which will give a bellows speed of 200 rpm. Cut a piece of cotton duck and tack it to the inside of A-frame to keep sand and dust from getting into the motor. Pieces of inner tube tacked over bearings will protect them from grit. Dry washing is a dusty operation, but it's a good way to get gold from those ancient deposits.

Good hunting!

Your reward—more than two ounces of nuggets!

Chapter 7
Recovering Fine Gold

One of the biggest problems facing the weekend prospector or amateur miner is separating the fine particles of gold from panned or "concentrated" gravel. Contrary to popular belief, small gold particles *do* float, making retrieval extremely difficult. When the prospector discovers many tiny flakes of gold in his pan mixed with thousands of grains of worthless sand, it is impractical to spend endless hours of tedious labor separating the gold. To the inexperienced prospector, it may seem his only alternative is to toss the whole mess back into the stream where he got it, but this is far from the case. With a few simple materials and a little know-how, that precious gold can be saved by a process known as *amalgamation.*

AMALGAMATION

Amalgamation with mercury is one of the oldest and simplest methods of recovering fine gold. The term "amalgamate" means to mix or blend, and that is exactly what happens when gold and mercury are brought together.

Mercury is a soft, liquid metal which tends to assimilate heavier metals such as gold and platinum while passing right over lighter materials such as fine gravel and black sand. But before gold can be amalgamated, it should first be cleaned of iron rust and other impurities. It is not unusual to find gold coated with iron stain and sulfides. These impurities make gold both very difficult to identify or amalgamate. During amalgamation, it is not uncommon for more than 30% of the rust-coated gold to be lost. Thus it becomes readily apparent that the first step is to clean the gold of these foreign particles.

CLEANING GOLD

There are numerous ways to clean gold, but one very simple

method employs nitric acid. First, place the concentrated gold-bearing sands into a plastic gold pan or other container that will not be affected by powerful acids. Cover the concentrates with about one-half inch of water and add a small amount of nitric acid until a slight "boiling" action occurs. This boiling begins when the ratio of water-to-acid reaches a concentration of about 10:1. In some cases no dilution is necessary; add straight acid.

Handling concentrated acids can be very dangerous because they have a tendency to suddenly "boil over" if water is poured into a container of acid.

Once you have the proper mixture in the plastic pan, swirl the liquid around, making sure all the material is completely covered. Then the acid solution can be washed away by dunking your pan in the stream. Pour off the excess water, and your concentrated sand and gold should now be ready for amalgamation.

MERCURY

It takes only a few drops of mercury to gather the gold in an average pan. A tiny quarter-pound vial of mercury can literally recover pounds of gold if used and cared for properly. The procedure is as follows: a small quantity of mercury is placed in the pan with the concentrates and agitated *under water.* The mercury should be worked thoroughly through the sand for complete contact with all the concentrates. The mercury will appear rough on the outside of the "ball" after it has gathered the gold. The roughness is due to the gold that has been assimilated.

At this point, the waste sands should be panned out, with the mercury-amalgam ball still remaining in the pan. Care should be taken not to lose any of the tiny drops of mercury which will gradually run together into a single mass.

During the amalgamation process, the ball gradually turns gray and becomes brittle. This mercury ball now contains all of the fine and powder gold that would otherwise have been nearly impossible to recover.

SEPARATING GOLD FROM THE MERCURY

There are a number of ways of recovering the gold from the amalgam ball, but the most common is retorting. The first step is to clean the amalgam by washing and then straining through a buckskin, chamois skin, or tight-weave canvas to recover as much mercury as possible. The remaining material is then boiled off to remove the mercury, either over a fire or with the use of concentrated acid.

THE RETORT

Since mercury is expensive, it may be reclaimed for use time and time again by capturing the fumes with the aid of a retort. While the retort is the most common method for reclaiming mercury, there are actually other ways the same job can be accomplished. We will discuss another of these ways later. The retort consists of a cast-iron pot with a tight-fitting lid. A tube connects the pot with a condenser which actually does the work of recovering the mercury.

The amalgam is placed inside the pot and heated. At a temperature of 675 degrees Fahrenheit, the mercury turns to gas and escapes through the tube to the condenser. The condenser is nothing more than a water-jacket wrapped around a tube. The water cools the mercury vapor as it passes through the tube, where it is condensed back into a liquid metal form. After all the mercury has been driven off, the pot is allowed to cool and the lid removed. All that remains in the pot is a mass of pure "sponge" gold.

THE BAKED POTATO METHOD

While the "retort" method is by far the most common way of freeing gold from the amalgam, it is not necessarily the most practical for the weekend prospector. Retorting requires a great deal of heat as well as a relatively complicated apparatus.

If there is only a small amount of amalgam and the prospector still wishes to reclaim his mercury, probably the most simple method is to scoop a small cavity in a potato. To do this, a large, well-rounded potato is first cut in half, then a small depression scooped into one side with a spoon or knife. The amalgam is poured into this cavity, the two halves of the potato placed back together as they originally were, and wired together. The potato is then wrapped in several layers of aluminum foil and placed in a campfire to "bake." In around 45 minutes (depending on the size of the potato and the amount of amalgam) the heat will have vaporized the mercury and driven it into the potato, leaving the gold in its free state in the

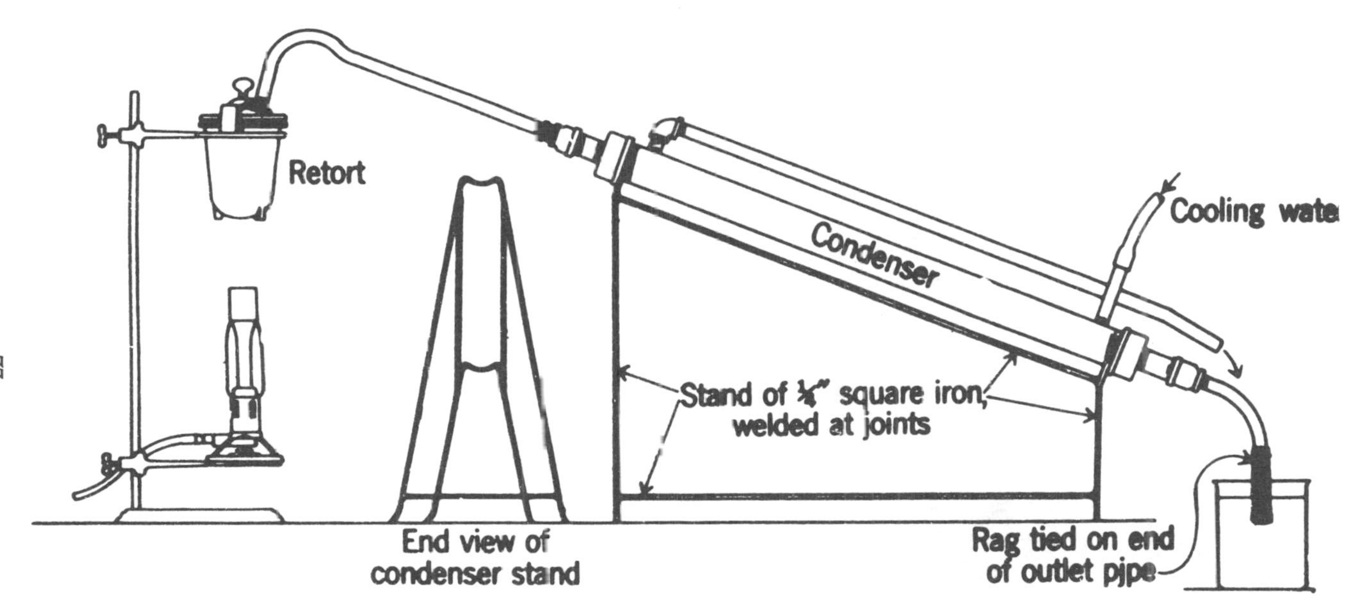

Diagram of a typical apparatus for retorting amalgam and quicksilver.
Courtesy of CALIFORNIA GEOLOGY

cavity. Once the gold is removed, the potato can be crushed, and the mercury panned out in an ordinary gold pan.

Caution should be observed when using this method, however. Mercury is extremely poisonous. DO NOT EAT THE POTATO! Care should be exercised, as occasionally the extreme heat will cause the potato to explode. The fumes from vaporized mercury can be deadly; always perform this process outdoors. Avoid contact with skin and open sores.

While it is true that mercury is expensive, I feel the most practical method for processing very small amounts of amalgam is to first recover as much of the mercury as possible by squeezing through a chamois skin as described earlier. Then add a few drops of *concentrated* nitric acid to the remaining amalgam, and the mercury will "boil" away, leaving the gold bright and shiny.

Recovered mercury can be used over and over, but as dirt begins to accumulate, the mercury will discolor and often break into many tiny particles, preventing complete amalgamation. When this happens, a 30:1 dilution of nitric acid will clean the dirt from the mercury, giving it a new lease on life. Both nitric acid and mercury can be obtained from your local prospecting supply store, or through many pharmaceutical and chemical supply houses. You will be amazed at the amount of fine gold that can be recovered through amalgamation that otherwise would have been lost!

Chapter 8
Metal Detectors and Dry Washing

In the past few years, there have been dramatic new improvements in metal detector technology. New detectors have been developed for one purpose only: finding gold. These detectors are primarily used for nugget hunting, but can be a tremendous aid to the prospector seeking the best spots to dry wash. Using a detector to locate placer deposits, rather than the time-consuming sampling method, can save hours and even days when searching for the right spot to set up your equipment.

Some of the well-known brands are manufactured by Fisher Electronics, Tesoro, White's Electronics and Compass. I have used a Fisher Electronics *Gold Bug* for a couple of years and can highly recommend it for any prospecting trip. I use it to locate the best-producing spots in the gold-bearing region I am working—and to find nuggets, as well.

When I find a location that shows a lot of fine gold, I set up my dry washer and work the deposit. This really speeds up recovery and guarantees some success. It is a lot of fun, and you can find some really good artifacts, too.

One of the many advantages of using a metal detector to look for gold is the fact that you will find nuggets—which bring a much better price than fine gold. For instance, if you are dredging, most of your gold will be very fine, which is worth 75% to 80% of the spot gold price, depending upon purity. Nuggets, however, command a higher price. If they have a nice shape or resemble something (like a bird, for instance), you might get as much as twice the gold content. Also, pieces of quartz with visible veins of gold in them bring high prices from mineral collectors.

Always use a headphone. Without headphones, you will miss a lot of small deep nuggets, especially where there is background noise, like wind blowing through the brush. A one-grain nugget will not make much of a sound; without headphones, you will usually miss it. Some patches have hundreds of detectable small nuggets, but almost none of any size. In a place like that, you would be wasting your time without headphones.

Even if it's lying on top of the ground, you're going to have to determine which of those pebbles is really a gold nugget. If you have to dig for it, you must determine which handful of dirt is pay dirt. But the worst part of it is that most of your targets won't be gold. They'll be nails, junk, hot rocks or what-have-you. The *only* way to be absolutely sure is to dig them up.

Just about any metal detector will sound off over a large nugget. The problem is that most nuggets are small. The *Gold Bug* will respond solidly to nuggets not much bigger than the head of a pin.

This is one of the most important tips I can give you. Use a small nugget, no larger than the size of a match head, as a tuning piece. It will be a big help when you are in bad ground and you're not sure whether your detector is tuned just right. Just put it on the ground occasionally and see if your detector sounds off over it. If not, your detector's ground adjustment is probably off; you should make the necessary adjustments until you can hear it.

Once you find a nugget-producing patch of ground, "grid" it to ensure that you cover every inch. Gridding simply means squaring off a patch and marking it with parallel lines no more than four feet apart. You can do this with a heavy tree limb or by dragging a piece of heavy chain on a rope back and forth. If the patch is very small, you should also rake it to remove surface rocks and other rubble; the closer you get your detector's coil to the ground, the more nuggets you'll find.

If you suspect that your target is on the surface, grab a handful of dirt and rocks and pass your hand across the top of the coil (make sure you're not wearing any rings or watches). If you get a response—but can't determine what in your hand is the target—put half the dirt in your other hand and check again. By repeating this process several times, you should be able to identify even very small targets.

Pinpointing is not difficult, even with the standard ten-inch coil. When you receive a positive sound, move the coil back and forth over the target to get an idea where your target is located. Next, loosen the soil with a pick and grab a handful of dirt. Recheck the hole with your coil. If the sound is still there, repeat. If the sound is gone, you should have the target in your hand. Pour a bit of dirt on top of the coil. If you hear nothing, pour it off and trickle some more onto the top of the coil. When the nugget hits the coil, you'll hear a distinctive "bleep." Simply blow the dirt away, and there is your nugget.

Sometimes little ferrous targets, like bits of wire, will give you fits. A nugget is easily pinpointed, but small ferrous targets will sound off over a wider area—and lots of needless digging will result. If you use your detector often, you will easily be able to tell the difference in sound between a nugget and a piece of ferrous trash. For the beginner who can't tell the difference—or the pro who likes to dig everything—an easy way of pinpointing elusive targets is to tilt the coil on its side so that just the small rounded edge moves over the ground. This only works on shallow targets, but it's a fast way to zero-in on small surface trash items.

Not only are most nuggets small, but often they are in highly mineralized soil. Mineralization may be so strong that it overloads the ground reject circuitry of many detectors and drastically reduces the performance of many others.

Many rocks have a higher iron content than the dirt they are found in. Since you adjust your detector to the surrounding ground, these rocks will sound off as a positive signal. They are quite plentiful in some areas, but you can easily deal with them. Listen very carefully to the sounds they make. They will all sound alike in the same patch, but a nugget among them will give a much sharper sound.

Some nugget hunters ground-balance over a hot rock until it no longer sounds off. This procedure may eliminate hot rock noises in a particular area, but it will also eliminate responses from small nuggets, since the detector has been tuned to the rocks and not the soil—which is where the nuggets are. Large pieces can be found in this manner, but don't forget that, for every good-sized nugget, there are many small bits. Don't pass them up; they are your bread and butter.

A hot rock will sound very strong at ground level, but when you raise the coil, the sound will rapidly fade. Most hot rocks cannot be detected at any distance.

In areas that are really littered with hot rocks, pay attention to the way they look. They are usually the same color in any given area: brown, red, black or gray. In an area where they are all red, for instance, check any rock that sounds off, but looks different—especially the ones that show quartz. It may not be just a hot rock, but a gold specimen.

When searching for nuggets in areas where there is a history of mining from the 1800s, I combine treasure hunting and relic collecting with my prospecting. I have found old hand-made tools and old bottles as well as gold and silver coins left behind or lost by miners.

A really good detector to use if you want to do the same is the Eagle Spectrum from White's Electronics. This unit is so easy to work with and does so many things, that it really makes getting out a lot of fun. It is completely computerized and as simple to operate as pushing a button. The display panel leads you step-by-step through the operating procedure, telling you what control to press and when to do it. It has four pre-set programs: prospecting, relic, coin and jewelry and beach—as well as two custom programs so that you can design your own.

Once you have selected a program like coin and jewelry, the display will identify each target hit and tell how deep it is buried (for instance, a quarter at six inches). The display can be back-lighted for easy reading on those days when it gets dark early or if you want to hunt at night. The prospecting program is a non-discriminating mode; all targets will produce an audio signal, but the display will show only a "VDI Number" for metals that could be gold. By digging only at those signals, most iron targets are eliminated.

When exploring old mining camps and ghost towns, I use the coin and jewelry mode. You'd be surprised at what turns up. White's also makes a great nugget shooting detector called the *Goldmaster II.* You might want to look at it for your prospecting outfit. It has some excellent features, and I've seen some nice nuggets found with it.

As you can see, there are many aids available to the modern prospector that only a few years ago seemed like a dream. Probably that is why we are seeing so many more successful prospectors and treasure hunters today. I find gold-hunting detectors to be very effective; when used properly, they are valuable tools that can pay for themselves.

Chapter 9
How to Stake a Claim

If, by chance, you should strike El Dorado and locate a rich deposit of gold, you will need to know how to claim it for your own. Remember that the first men involved in writing California laws were either miners or men connected with the Gold Rush in some way. Men like Hearst, Armour and Stanford, whom we associate with newspapers, meat packing, and universities, began their fortunes in the gold fields. It was these same men who helped draft the 1872 Mining Laws.

The states were dependent on the gold-mining industry, and its land laws still favor the miner. If you own your home, you probably are aware that you do not always get mineral rights with your title. Mineral rights have to be claimed and proved if contested. Mineral rights belong to the locator.

If you locate a deposit you feel is worth developing, there are some basic things you can do to protect your discovery. First, as you know, there are two types of gold mines: placer and lode. Each has its own set of rules and regulations.

A placer claim may consist of as much as twenty acres for each person signing the location notice. Placer gold, again, is free gold found in gravels and alluvial deposits. You must place a Notice of Location on a post, tree, rock in place, or on a rock monument you build, showing the name of the claim, who is locating it, the date, and the amount of area claimed. You must mark the boundaries of your claim so that it may be easily traced. You must also identify the location of your claim by reference to some local landmark or natural object (such as a stream or rock formation) or a permanent monument. Finally, you must file a copy of the Location Notice with the local county recorder within ninety days of the date of location.

Claim marker. Courtesy of Allied Services

A lode claim uses a different procedure. Lode gold, again, is gold in a vein, which must be mined and separated from its mother rock. The claim may consist of as much as 1500 feet along the sidelines of the vein, and 300 feet on each side of the middle of the vein. Within ninety days of the location of any lode-mining claim, you must place a post or stone monument at each corner of the claim. The posts must be at least four inches in diameter, and the stone monuments must be at least eighteen inches high. A *Notice of Location* form must be posted on the claim, and a lode Location Notice filed with the county recorder within ninety days.

This section is not intended to be a complete thesis on all the rules and regulations regarding mining claims, but it will help to protect you in the initial phases should you make a rich discovery.

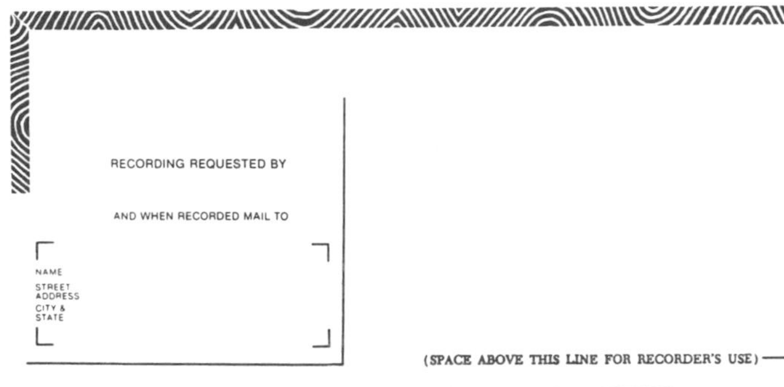
RECORDING REQUESTED BY

AND WHEN RECORDED MAIL TO

NAME
STREET ADDRESS
CITY & STATE

(SPACE ABOVE THIS LINE FOR RECORDER'S USE)

PLACER MINING CLAIM LOCATION NOTICE

TO WHOM IT MAY CONCERN. Please take notice that:

1. The name of this claim is ______________________, a placer mining claim.

2. This claim is situated in:

NE ¼ ☐ NW ¼ ☐ SW ¼ ☐ SE ¼ ☐ Sec. ______ T. ______ R. ______ Mer. ______

NE ¼ ☐ NW ¼ ☐ SW ¼ ☐ SE ¼ ☐ Sec. ______ T. ______ R. ______ Mer. ______

NE ¼ ☐ NW ¼ ☐ SW ¼ ☐ SE ¼ ☐ Sec. ______ T. ______ R. ______ Mer. ______

NE ¼ ☐ NW ¼ ☐ SW ¼ ☐ SE ¼ ☐ Sec. ______ T. ______ R. ______ Mer. ______

in the ______________ Mining District, County of ______________, State of California.

The acreage claimed is ______________ acres.

3. The date of this location is the ________ day of ______________, 19____ on which date the notice of location was posted on the claim.

4. The locator or locators of this claim are:

Name(s)	Current Mailing or Residence Address

5. If claim cannot be described by quarter-section, the boundaries of the claims and the land taken are described as follows:

Commencing at the discovery monument where this notice is posted, thence ________ (direction) to the ________ corner which is the point of the beginning to describe the boundaries,

thence ________ (direction) ________ feet to the ________ corner,

thence ________ (direction) ________ feet to the ________ corner

thence ________ (direction) ________ feet to the ________ corner,

thence ________ (direction) ________ feet to the point of beginning.

6. The discovery monument is situated at the point of discovery about ______________ (distance from natural object or permanent monument and give direction as accurately as possible to identify the claim located)

SIGNATURE(S)

______________ ______________

______________ ______________

______________ ______________

______________ ______________

CLAIMS LOCATED AFTER OCTOBER 21, 1976, MUST BE RECORDED WITH THE BUREAU OF LAND MANAGEMENT **WITHIN 90 DAYS AFTER DATE OF LOCATION.**

CSO 3800-1
6/79

WOLCOTTS FORM 1130 – PLACER MINING CLAIM LOCATION NOTICE – Rev. 3-80

For more details, you may be interested in obtaining a copy of STAKE YOUR CLAIM, by Mark Silva, which explains the legal and business aspects of gold prospecting and claims. Another source of information is the state office of the Bureau of Land Management, for their own publications.

ASSAYING AND REFINING

If you think you have found a good gold spot, you'll want to find out its real value. Take your samples to an assayer. He will give you a full report on your sample. Who knows, you might have also located a silver mine as well!

If your assay shows a good percentage of gold, you have several options. You can contact a mineral resource company to help you develop the lode, or you can develop the mine yourself. In the latter case, you'll ship the material to a smelter to be refined and returned to you. The refinery may buy it directly from you as well. If you're successful at placering, you can sell your gold directly to a refinery, jeweler, dentist, assayer, or prospectors' supply store.

Check the paper for the current price of gold before approaching a possible buyer. Your unrefined gold will sell for less, naturally, than the refined gold price quoted in the paper; however, the daily market price will give you a ball-park figure to start from. Keep in mind that any fair-size nugget will be worth more than its weight. This is due to the fact that they are sought after as specimens by collectors and by jewelers for rings and other jewelry.

Now you know where to look, how to find gold and what to do with it, so come on out and join us. I'll be looking for you.

JIM KLEIN

Chapter 10
The Most-asked Questions About Dry Washing

1. HOW DOES A DRY WASHER WORK?

The principle of a dry washer is the same as any gold separation device. The idea is to suspend the gravel enough to allow the heavy particles to settle out of suspension into a collector box or sluice while the lighter particles (of no value) can flow away. The most effective method of suspension for dry processing is air. A sufficient air flow to suspend gravel will save a good percentage of values, but this in itself is not enough. Vibration and oscillation properly induced has proven to be essential for fine gold recovery.

2. HOW DOES "ELECTROSTATIC CONCENTRATION" WORK ?

Dry air (preferably heated), generates a positive electrostatic field as it passes through the cloth at the bottom of the riffle section. This positive charge has the capability to hold metal or any conductive material, much like a magnet.

3. WHAT ARE THE FINEST PARTICLES OF GOLD A DRY WASHER CAN SAVE?

A properly designed dry washer can be one of the most effective gold recovery systems. If properly operated, gold as fine as 200 mesh can be consistently recovered. (200 mesh is gold that can pass through a screen with 200 holes per inch, about the size of talcum powder.)

4. WHERE ARE THE BEST AREAS TO LOOK FOR GOLD WITH A DRY WASHER?

Check either with your local prospector store, book store, local department of geology or mining equipment supplier for books or maps describing gold locations. There are many published booklets and maps on gold prospecting, and it should not be difficult to find a few.

5. CAN A DRY WASHER RECOVER GOLD FROM DAMP SAND?

Damp sand will not separate into individual particles for direct contact with any recovery system. The damp sand may hold values in clumps and clods and make recovery of values impossible. Also, damp sand cannot receive an electrostatic charge. Most dry washers with an air blower will dry some of the sand, but the material should be run through more than once in order to maximize recovery.

6. HOW MUCH MATERIAL CAN A MAN PROCESS THROUGH A DRY WASHER?

The average capacity of a man is about one ton per hour. This figure depends on the elements involved such as heat, altitude, gravel conditions and the strength and endurance of the person doing the shoveling.

7. WILL A DRY WASHER RECOVER LARGE NUGGETS AS WELL AS FINE GOLD?

Most units now have a nugget trap built into the classifier screen. Although the likelihood of finding a large nugget that will not pass through the 1/2 inch classifying screen is rare, it is possible and any processor should have such a safety valve. A nugget larger than the holes in the 1/2-inch classifying screen will be easily spotted by most prospectors and not lost. Believe me, if one that large shows up, you'll see it. You may want to run your material from the screen a second time to be sure.

8. WHY IS AN ADJUSTABLE VIBRATOR IMPORTANT TO A DRY WASHER?

Many dry washers on the market today have vibrators, but few have adjustable oscillation, which is extremely important because all gravel is not the same consistency, nor are the values you expect to recover. An adjustment may be necessary where varied ground conditions may occur. The wrong oscillation could conceivably cause the gold to move out and only the gravel recovered!

9. HOW OFTEN DOES A DRY WASHER HAVE TO BE CLEANED OUT?
Clean-up time varies with the amount of heavy concentrate that builds up in the riffle tray. When a heavy concentration of black sand is prevalent, clean-up can be necessary as often as every thirty minutes. The normal clean-up time in an average condition would be once every two hours.

10. CAN I SUCCESSFULLY MAKE ONE MYSELF?
You can often find instructions on "how to build," but some instructions are obsolete and do not include up-to-date information for an adequate system. One might consider purchasing some of the component parts to build a dry washer less expensively. Items such as the electrostatic collector tray and the vibration device would be extremely difficult to duplicate. Often, home-built units are heavier and can suffer in performance.

11. WHAT IS THE BEST ANGLE FOR THE RECOVERY BOX?
The optimum angle drop of the recovery box is about seven inches. The angle must be adjusted at the location due to varying gravel conditions. The proper angle of the recovery tray cannot be determined until a continuous and consistent flow of gravel is passing over the tray.

12. SHOULD THE CONCENTRATOR COLLECT BLACK SAND?
The sluice should collect all of the gold-bearing black sand. In some areas, black sand is abundant and not always gold-bearing. The sluice is designed to hold the heaviest particles that would be available in any particular area. For example, some have been successful in collecting sheelite, with a specific gravity as low as five, much lower than the average weight of common black sands.

13. IF THE COLLECTOR BOX IS LOSING BLACK SAND, WHAT SHOULD YOU DO?
The first thing to consider is, is the box actually losing values? It may appear that value is being lost, only to find that after checking the tailings that little or no values existed. If the entire area is noth-

ing but heavy black sand, try slowing down the feed somewhat to allow for a more selective concentration.

14. WHAT IS THE SIMPLEST TEST FOR RECOVERY?
Let the machine run without gravel intake for approximately three minutes and simply observe the front end of the sluice for values. Often a small paintbrush will brush away some of the excess particles to see gold or other values at the bottom of the riffle.

15. CAN A DRY WASHER HANDLE ONE PERSON SHOVELING AT FULL SPEED?
Yes, but an average person cannot shovel at full speed for any length of time. Some larger units process the work of two persons shoveling at the rate of approximately two tons of bank run per hour.

16. WHAT IS THE BEST POSITION FOR THE ADJUSTABLE COUNTERWEIGHT ON THE VIBRATOR, IN VARYING CONDITIONS?
Most adjustable counterweights are pre-set at the factory for average conditions. If the gravel contains moisture, increase the vibration to a higher frequency, which will separate damp gravel at a faster rate. For dryer conditions, decrease vibration speed. For normal operations, a slower frequency will provide best results.

17. WHAT IS THE OPTIMUM POSITION FOR THE CLASSIFYING HOPPER?
The hopper should be run as flat as possible, maintaining a minimum angle that will allow oversize material to run off. Try the center hole position to start.

18. WHAT MAINTENANCE IS REQUIRED FOR A DRY WASHER?
The only required maintenance is to change the engine oil after every 24 hours of operation.

Glossary

ALLOY. A solid solution of two or more minerals.

ALLUVIAL. Loose gravel, soil, or mud, deposited by water.

AMALGAM. Normally a physical alloy of mercury with gold or silver.

ARRASTRE. A circle of stones where ore was crushed during the early days of gold mining; a primitive but effective method of separating gold from quartz.

ASSAY. To evaluate the quantity and quality of minerals in an ore.

AURIFEROUS. Containing or bearing gold.

BAR. A name given to sandbars and rock and gravel deposits found in rivers, primarily when they are gold-bearing.

BENCHES. Rock or gravel shaped like terraces or steps. Bench placers are found on the canyon walls above present stream beds.

BLACK SAND. Heavy minerals, typically made up of magnetite, tourmaline, ilmenite, chromite and cassiterite. They are found in rivers, beaches, sluice boxes and pans. Black sand will usually be found with gold, but gold is not always found with black sand.

COLOR. Any amount of gold found in a prospector's pan after a sample of dirt has been panned.

DIGGINGS. A claim or place being worked.

DIORITE. A granular, crystalline, igneous rock in which gold sometimes occurs.

DREDGING. A method of vacuuming gold-bearing gravels from river or stream bottoms.

DRIFT. A horizontal tunnel following a vein or gold-bearing gravels.

DRY WASHER. A machine which separates gold from gravels by the flow of forced air.

FLOAT. Loose pieces of ore broken off a vein outcropping. Prospectors will follow the float to its source to locate a lode.

GLORY HOLE. A small but very rich deposit of gold ore.

GRAVEL BENCHES. Gravel deposits left on canyon walls through stream erosion.

GULCH. A small canyon or ravine.

HARDROCK MINING. Another term for lode mining.

HEAD FRAME. The support structure located at the entrance of a mine over a shaft. Used for hoisting.

HYDRAULIC MINING. A very destructive and now-outlawed form of gold mining used during the Gold Rush. Giant hoses were used to force great streams of water onto canyon walls containing gold-bearing gravels. The walls were washed away into sluice boxes, where the gold was then picked out.

IRON PYRITE. A common mineral consisting of iron disulfide, which has a pale, brass-yellow color and a brilliant metallic luster. Also called fool's gold.

LODE. A vein of gold mined either through a tunnel or a shaft.

MATRIX. The material in which the gold is found.

PLACER. Free-occurring gold which is usually found in stream and river gravels.

POCKET. A rich deposit of gold occurring in a vein or in gravels.

POKE. A leather pouch used by old-time miners to hold their gold.

QUARTZ. A common mineral, consisting of silicon dioxide, that often contains gold or silver.

RETORT. A device used to separate gold from mercury.

RICH FLOAT. Gold-bearing rocks worked loose from a lode.

ROCKER. A device used by the early miners during the Gold Rush. This was a sluice box mounted on rockers with a hopper on the top to classify the material. Gravels were shoveled into the hopper; then water was poured on top, washing the gold-bearing material down over the riffles while the hopper was rocked. The rocking helped the gold to settle.

SCHIST. A crystalline rock which is easily split apart.

SLUICE BOX. A trough used in continuously moving water with obstructions to trap gold.

STAMP MILL. A machine used to crush ore.

SULFIDE. A compound of sulfur and any other metal.

TAILINGS. The material thrown out after ore is processed. The tailings from the early mines, where the miners were sometimes very careless, have produced significant amounts of gold and other valuable minerals.

TERRACE DEPOSITS. Gravel benches high on canyon walls.

WALL. The rock on either side of a vein.

WIRE GOLD. Gold thinly laced through rock.